ENGINEERING PRACTICE IN THE UNITED STATES

Engineering Graduate Education and Research

Panel on Engineering Graduate Education
  and Research

Subcommittee on Engineering Educational Systems

Committee on the Education and Utilization
  of the Engineer

Commission on Engineering and
  Technical Systems

National Research Council

NATIONAL ACADEMY PRESS
Washington, D.C. 1985

NATIONAL ACADEMY PRESS • 2101 Constitution Ave., NW • Washington, DC 20418

NOTICE: The project that is the subject of this report was approved by the Governing Board of the National Research Council, whose members are drawn from the councils of the National Academy of Sciences, the National Academy of Engineering, and the Institute of Medicine. The members of the committee responsible for the report were chosen for their special competences and with regard for appropriate balance.

This report has been reviewed by a group other than the authors according to procedures approved by a Report Review Committee consisting of members of the National Academy of Sciences, the National Academy of Engineering, and the Institute of Medicine.

The National Research Council was established by the National Academy of Sciences in 1916 to associate the broad community of science and technology with the Academy's purposes of furthering knowledge and of advising the federal government. The Council operates in accordance with general policies determined by the Academy under the authority of its congressional charter of 1863, which establishes the Academy as a private, nonprofit, self-governing membership corporation. The Council has become the principal operating agency of both the National Academy of Sciences and the National Academy of Engineering in the conduct of their services to the government, the public, and the scientific and engineering communities. It is administered jointly by both Academies and the Institute of Medicine. The National Academy of Engineering and the Institute of Medicine were established in 1964 and 1970, respectively, under the charter of the National Academy of Sciences.

Support for this work has been provided by the National Science Foundation, the Department of the Air Force, the Department of the Army, the Department of Energy, the Department of the Navy, and the National Aeronautics and Space Administration. Additionally, assistance has been provided through grants from the Eastman Kodak Company, Exxon Corporation, the General Electric Company, the IBM Corporation, the Lockheed Corporation, the Monsanto Company, and the Sloan Foundation.

Library of Congress Catalog Card Number 85-71643

ISBN 0-309-03549-X

Printed in the United States of America

Preface

This panel report was prepared as part of the overall study of engineering education and practice in the United States that was conducted under the guidance of the National Research Council Committee on the Education and Utilization of the Engineer. Many of the findings and recommendations of this report were included in the summary report of the committee,* but it was possible to address the various topics in more detail here.

The basic objective of the Panel on Engineering Graduate Education and Research was to examine the state of engineering graduate study in the United States, to define what its future should be, and to discuss its relationship to research. The panel concluded that graduate study has become steadily more important to the engineering profession in recent years. However, no fundamental change in the structure of engineering graduate education appears to be needed at this time.

The study focuses principally on increasing the supply of highly qualified doctoral recipients who are U.S. citizens, particularly with respect to academic employment. It also gives attention to the importance of master's-level work and to the need for access to part-time programs for engineers who are employed full time. The report includes recommendations of ways to provide the financial resources

---

* *Engineering Education and Practice in the United States: Foundations of Our Techno-Economic Future* (Washington, D.C.: National Academy Press, 1985).

to support a greater number of doctoral students. Most of the support would have to come from government sources, but increased support from industry is encouraged. The study also makes certain recommendations that would help maximize the effectiveness of industry–university cooperation.

The panel believes that the information contained in this study will be of help to those interested in dealing with the nature and scope of engineering graduate study. We especially hope that our recommendations will encourage the nation to take the necessary measures to sustain the present strength of our engineering educational establishment and to help it grow stronger in the future.

John D. Kemper
Chairman

Panel on Engineering Graduate Education and Research

JOHN D. KEMPER, Chairman; Professor of Mechanical Engineering, University of California, Davis
JAMES F. MATHIS, Vice-President, Exxon Corporation
IRENE C. PEDEN, Professor of Electrical Engineering, University of Washington at Seattle
JOSEPH E. ROWE, Vice-Chairman and Chief Technical Officer, Gould, Inc.
LELAND J. WALKER, Chairman, Northern Engineering and Testing

Committee on the Education and Utilization of the Engineer

JERRIER A. HADDAD, Chairman, (IBM, Ret.)
GEORGE S. ANSELL, Dean of Engineering, Rensselaer Polytechnic Institute (now President, Colorado School of Mines)
JORDAN J. BARUCH, President, Jordan J. Baruch Associates
ERICH BLOCH, Vice-President, IBM Corporation (now Director, National Science Foundation)
DENNIS CHAMOT, Associate Director, Department for Professional Employees, AFL/CIO
EDMUND T. CRANCH, President, Worcester Polytechnic Institute
DANIEL C. DRUCKER, Dean of Engineering, University of Illinois at Urbana (now Graduate Research Professor of Engineering Sciences, University of Florida at Gainesville)
FRED W. GARRY, Vice-President, Corporate Engineering and Manufacturing, General Electric Company
JOHN W. GEILS, Director of AAES/ASEE Faculty Shortage Project (AT&T, Ret.)
AARON J. GELLMAN, President, Gellman Research Associates, Inc.
HELEN GOULDNER, Dean, College of Arts and Science, Professor of Sociology, University of Delaware
JOHN D. KEMPER, Professor, Department of Mechanical Engineering, University of California at Davis
EDWARD T. KIRKPATRICK, President, Wentworth Institute of Technology

COMMITTEE MEMBERS

ERNEST S. KUH, Professor of Electrical Engineering and Computer Science, University of California at Berkeley

W. EDWARD LEAR, Executive Director, American Society for Engineering Education

LAWRENCE M. MEAD, JR., Senior Management Consultant (Senior Vice-President, Ret.), Grumman Aerospace Corporation

M. EUGENE MERCHANT, Principal Scientist, Manufacturing Research, Cincinnati Milacron, Inc. (now Director, Advanced Manufacturing Research, Metcut Research Associates, Inc.)

RICHARD J. REDPATH, Vice-President, Ralston Purina Company

FRANCIS E. REESE, Senior Vice-President, Monsanto (now retired)

ROBERT M. SAUNDERS, Professor, School of Engineering, University of California at Irvine (Chairman, Board of Governors, AAES, 1983)

CHARLES E. SCHAFFNER, Executive Vice-President, Syska & Hennessy

JUDITH A. SCHWAN, Assistant Director, Research Labs, Eastman Kodak Company

HAROLD T. SHAPIRO, President, University of Michigan

MORRIS A. STEINBERG, Vice-President, Science, Lockheed Corporation

DONALD G. WEINERT, Executive Director, National Society of Professional Engineers

SHEILA E. WIDNALL, Professor of Aeronautics and Astronautics, Massachusetts Institute of Technology

Staff

WILLIAM H. MICHAEL, JR., Executive Director
VERNON H. MILES, Staff Officer
AMY JANIK, Administrative Assistant
COURTLAND S. LEWIS, Consultant

Government Liaison

LEWIS G. MAYFIELD, Head, Office of Interdisciplinary Research, National Science Foundation

Contents

Executive Summary 1

1. The Challenge 4

2. The Background 8
    Technological Developments During World War II, 8
    Grinter Report, 1952–1955, 11
    PSAC Report, 1962, 12
    Goals Study, 1963–1968, 12
    Retrenchment, 1969–1971, 15

3. Supply and Demand 17
    Supply of Ph.D.s, 18
    Quality of Entering Graduate Students, 23
    Ph.D.s in Academic Employment, 26
    Ph.D.s in Industry, 37
    Increasing the Supply of Ph.D.s, 46
    Educational Technology and Productivity, 51
    Findings and Recommendations, 55

4. Women and Minorities in Engineering 58
    Women in Academic Careers, 62
    Finding and Recommendation, 71

5. The Master's Degree 72
   Findings and Recommendations, 82

6. The Doctor's Degree 83
   Findings and Recommendations, 92

7. Nontraditional Graduate Programs 93
   The Doctor of Engineering, 94
   The Engineer Degree, 96
   Finding and Recommendation, 97

8. The Engineering Faculty 98
   The Need for the Doctor's Degree, 99
   Adjunct and Part-Time Faculty, 100
   Faculty Development and Self-Renewal, 101
   Findings and Recommendations, 103

9. University–Industry Interactions .104
   The Appropriate Nature of Industrially Sponsored
      Research, 106
   Patents, 108
   Consulting, 110
   Faculty Conflicts of Interest, 116
   Findings and Recommendations, 116

   Notes 118

Executive Summary

The Panel on Engineering Graduate Education and Research prepared this report as a part of the overall effort of the National Research Council's Committee on the Education and Utilization of the Engineer. Following is a summary of the major topics covered in this report.

The challenges to engineering education, and especially to graduate education, are extraordinary. It is unlikely that any one university will be able to offer a menu of graduate education in every discipline and subdiscipline of engineering, largely because of the enormous expense of laboratory research facilities. Some large institutions will come close to the ideal of being able to offer graduate education in all areas of engineering, but most will have to be selective.

The growth in graduate engineering education has been largely a phenomenon of the post–World War II era. A strong influence toward postbaccalaureate education was the realization that the solution of many postwar problems would depend on more science and mathematics in engineering curricula and on larger numbers of engineers educated at advanced-degree levels.

Although it is difficult to predict the demand for engineers, numbers of future graduates, and thus the supply of engineers, can be predicted from current enrollments. On this basis, the nation can probably expect 3,800 to 4,000 engineering Ph.D.s per year by 1988, with approximately 40 percent of this total being foreign nationals on temporary visas. About one-third of the total is expected to enter academic employment. However, even though the total number of Ph.D.s is increasing, the actual number available for academic employment each year is expected to average only about 100 more per year than was the case during the 1970s. This is perceived to be a serious situation, since the current shortage of engineering faculty developed during the 1970s. (In 1982 an estimated 1,400 engineering faculty positions nationwide were unfilled; in 1983 this number went up to 1,570.) Simultaneously, the student:faculty ratio for engineering schools increased by 37 percent between 1976 and 1982. If enough faculty positions were allocated to maintain student:faculty ratios at the 1976 level, 6,700 new faculty would be needed.

To help alleviate these problems, the Panel on Engineering Graduate Education and Research recommends the following:

• Universities should seek to reduce current high workloads, improve salaries, and provide state-of-the-art facilities for instruction and research.

• The number of engineering doctoral fellowships should be increased so that a greater proportion of U.S. citizens from the top decile of B.S. degree programs will be encouraged to enter doctoral study; about 1,000 new "starts" should be available each year, with stipends at least equal to 50 percent of industrial starting salaries; industry and government should work together on this program, which is estimated to cost between $60 million and $70 million per year. A concomitant increase in research funding of $200 million per year for academic institutions is needed.

• Increased emphasis on engineering research within the National Science Foundation is strongly encouraged, although it is recognized that other agencies will also continue to be strong supporters of engineering research.

• The available base of facilities and equipment has fallen far below what is needed. The recommended increase in the number of Ph.D.s, plus overcoming the full 6,700 "shortfall" in faculty, would require a one-time nationwide investment of $450 million to $1 billion for new facilities, depending upon their sophistication.

The representation of women in engineering seems to be increasing at all academic levels. However, this is not the case for minority groups. The lack of minorities in graduate school is perceived to be principally a "pipeline" problem. Major efforts are needed with respect to minority groups at the junior high and high school levels, and upgraded retention programs are needed at the college level. Efforts must be made to eliminate discrimination, real or perceived.

EXECUTIVE SUMMARY

The master's degree is becoming the preferred entry degree for professional practice in some fields of engineering, but cannot yet be universally regarded as the entry-level degree. While well-supervised master's degrees have great value for students, they should not be allowed to degenerate into routine exercises. Master's-level study has been shown to help forestall becoming technically outdated, and employers should provide opportunities for continuing education, both degree-oriented and non-degree-oriented. Courses offered on television provide advantages in this regard.

Most engineering faculty should have the doctorate, particularly in research universities. However, schools that emphasize programs in engineering technology have not demanded doctorates for their faculty, nor have schools that specialize in undergraduate programs. Even research universities effectively utilize a small percentage of nondoctoral faculty by employing engineers in professional practice to enrich the applications aspect of their programs.

Industry appears to be showing more interest in employing engineering Ph.D.s, although it is not possible to make a definitive forecast of demand. In 1981 a survey of the National Research Council showed a zero unemployment rate for 1980 engineering Ph.D. graduates. Thus, the only clear indication of Ph.D. shortage is the one related to unfilled faculty positions. While it cannot be shown that there is a shortage of engineering Ph.D.s for industry, there does not appear to be a surplus either.

Closer ties between industry and engineering education should be fostered, bearing in mind the differing purposes of these two groups. Industrially sponsored research in universities should be free of secrecy constraints and should be as general as possible so that students' learning experiences can be transferred to a variety of future needs. Outside consulting by faculty members should be encouraged, provided it is of the type that supports and helps to improve the academic programs of the university. Faculty should scrupulously avoid conflicts of interest—a situation, for example, in which a faculty member is a principal investigator within a university on a project funded by a company for which the faculty member is simultaneously an officer, director, or significant owner.

1
The Challenge

The challenge confronting engineering and engineering education has been best expressed in a 1983 policy statement by the National Science Board:

> The United States is at a critical juncture in its industrial leadership. Not since Sputnik in 1957 has there been so much cause for concern about the adequacy of our science and technology base and our ability to capitalize on our scientific strengths to sustain industrial leadership. We face foreign competitors who have growing skills, lower costs, and higher productivity growth. These factors affect the security of our Nation, the standard of living of our people, and our legacy for future generations. [Statement on the Engineering Mission of the NSF Over the Next Decade as Adopted by the National Science Board at Its 246th Meeting on August 18–19, 1983.]

With the appearance of this external challenge, a challenge of another kind was manifesting itself, namely, the accelerating pace of technological development, which poses new problems vis-à-vis the appropriate education of scientists and engineers, especially at the graduate level. Rapid progress in computers has probably been the most important development with regard to engineering in the last decade, because the expansion of their power has been accompanied by a remarkable reduction in size and cost, significantly changing the manner in which engineering is practiced. Problems that formerly could not be solved because of their complexity or nonlinearity are now readily treated with computers. Massive systems problems, previously intractable because of the amounts of information to be processed, are now managed routinely, often in real time. Methods of laboratory experimentation have been drastically changed because data processing occurs as the experiments are run. The entire design and manufacturing process is being transformed into a substantially automated activity, and methods of conducting business are being revolutionized by the advent of electronic mail.

Interestingly, the rapid advance in computer capabilities has also stimulated experimental research, because our ability to model natural phenomena on a computer has in some cases outrun our knowledge of nature itself. In a very real sense, developments in computer power and in experimental research move in parallel, each stimulated by the other.

The technological challenge embraces virtually every field of engineering knowledge. In electrical engineering, particularly with regard to solid-state electronics and very large scale integrated (VLSI) circuits, computer-aided design (CAD), robotics, artificial intelligence, and other computer-related specialties, universities have found it difficult to keep up with the pace of development, primarily because of the cost of equipment and facilities and the difficulty of finding qualified new faculty.

At present it does not appear likely that any given institution can expect to offer a menu of graduate education in all subdisciplines of electrical/computer engineering. A few schools have already emerged as centers of excellence in such areas as robotics, remote sensing, microwave techniques, and microelectronics. This may be a trend of the future, and it may be a corollary that a department strongly identified with a single or a few subdisciplines will have to relinquish the possibility of offering instruction and/or research facilities in other areas that might be attractive to graduate students. Nevertheless, these are fields in which universities with engineering graduate programs must participate, in varying degrees. New developments in communications, remote sensing, imaging, optics, and the submillimeter portions of the electromagnetic spectrum also pose challenges for electrical engineering.

In civil engineering, the status of our entire system of constructed works represents an urgent need, particularly because of the deterioration of the nation's transportation, water, and waste systems. The problems of design to protect against such natural hazards as earthquakes, wind, and rain also need further work, even though they are not new. Accommodation of hazardous waste products, better utilization of existing materials, and development of new materials for construction are other challenges to civil engineering.

The development of ceramic composites and fibers and the need for new kinds of materials in keeping with resources available pose major challenges to the field of materials science. At the present time the development of polymers for biomaterials often has no natural home within the academic community, with the result that it is difficult to attract good faculty and Ph.D. students to such research programs.

The outlook in chemical engineering is for continued development of process technologies. Scientific advances are being made in many areas of importance, including analytical techniques, mathematical modeling, materials, information technology, biology, and catalysis. Reducing energy consumption and waste emissions will continue to be important goals that require new process technologies. The reduction to practice of recombinant DNA technology for new drugs and agricultural chemicals is an important development that in turn will generate new challenges.

In mechanical engineering, the fields of combustion, heat transfer, theoretical aspects of turbulence, and computational fluid dynamics need further research and study. The advent of new materials calls for the development of new theoretical understanding of large deformations and damage assessment of these materials. New instrumentation and large-scale computation facilities are expensive but vital to these fields.

Large-system simulation has become an important tool in nuclear engineering research and education. Probabilistic risk assessment calls for sophisticated computer modeling. Nuclear engineers are presently dealing with the statistics of highly improbable events; this implies a kind of mathematics not found on the conventional graduate student course menu. Nuclear engineering education faces the additional problems of scarcity of graduate students who are U.S. citizens, difficulty in attracting young people into the field, and difficulty in obtaining new faculty.

In agricultural engineering there is an increasing need for computers because of the importance of large data-handling systems such as National Oceanic and Atmospheric Administration (NOAA) weather tapes, soil base data, and geological maps of underground water supplies. Overall agricultural system design is a new challenge because of the advent of microprocessors for controlling machinery and other equipment.

Graduate programs in minerals are generally faced with problems in acquiring state-of-the-art facilities for faculty and student use. In the fields related to exploration, graduate programs are moving in the direction of computer-assisted experimental work and modern airborne remote-sensing techniques. High-resolution seismic work to locate underground cavities is expensive and sophisticated, severely limiting the ability of academic institutions to educate advanced students.

Several common themes emerge from the foregoing descriptions of new areas in important engineering disciplines and their impact on graduate education, namely:

- the importance of large-scale computation and the resultant problems for academic institutions,
- rapid advances in available (usually expensive) instrumentation and its importance for experimental work and model validation, and
- the significance of interdisciplinary research and instruction at the forefront of many of these fields.

Funding for computers and instrumentation must be found, whether through government or private sources, and in some fields only regional or national entities can possibly address the problem. Corporate consortia are possibilities for other fields.

It may be desirable, under these circumstances, to consider the development of graduate programs in multidisciplinary or interdisciplinary fields, since many of today's engineering problems require such an approach. Bioengineering, materials engineering, environmental engineering, and manufacturing engineering are examples of fields that involve significant input from several engineering and scientific disciplines. However, many academic units lack the resources to establish new programs in these fields. Furthermore, it is not possible for all universities to be outstanding in all fields, and most will choose to focus their resources on a selected group of fields commensurate with their resources and to develop excellence in those areas. An additional problem is that young faculty attempting to work in new interdisciplinary fields may find their career advancement at risk if their more senior colleagues do not accept the new, unfamiliar work as meeting the academic standards of established fields.

This, then, represents the environment within which engineering graduate education and research must develop in the future. Ways will have to be found for new programs to develop and flourish within existing academic organizations and for new organizational structures to develop when needed. Graduate study will have to be made more attractive so that enough bright young men and women will be available to provide faculty for the nation's engineering schools. Resources for buildings and equipment will have to be provided so that sufficient numbers of well-educated new engineers can be provided to meet the future needs of our country.

2
The Background

The story of graduate education in engineering belongs largely to the post–World War II era. Just before the war the number of engineering master's degrees per year in the United States was only about a thousand. Doctor's degrees averaged about a hundred per year.* By 1949 these numbers had quadrupled, and by the 1970s the number of master's degrees had increased 15-fold, and doctor's degrees, 30-fold. (See Table 1; Table 2 gives distribution by field for 1982–1983.)

Several events have had major impacts on the evolution of graduate engineering education. Among them are the technological explosion of World War II, the publication of the Grinter Report in 1955, the publication of the President's Science Advisory Committee (PSAC) report in 1962, the Goals Study of 1963–1968, and the retrenchment of 1969–1971. Each of these is discussed below.

Technological Developments During World War II

During World War II, the public was dazzled by a succession of technological marvels. The world entered the war without such developments, for example, as jet aircraft, effective radar, and atomic energy, and it emerged with them. The example of radar is of special interest.

---

* Of the engineering baccalaureates awarded in 1940, the master's degrees were 12 percent and the doctor's degrees were 1.5 percent.

TABLE 1 U.S. Engineering Degrees, 1950–1983

Year Ending	Bachelor's Degrees		Master's Degrees		Doctor's Degrees	
	Foreign Nationals	Total	Foreign Nationals	Total	Foreign Nationals	Total
1950	n/a	48,160	n/a	4,865	n/a	492
1951	n/a	37,887	n/a	5,134	n/a	586
1952	n/a	27,155	n/a	4,132	n/a	586
1953	n/a	24,165	n/a	3,636	n/a	592
1954	n/a	22,236	n/a	4,078	n/a	590
1955	n/a	22,589	n/a	4,379	n/a	599
1956	n/a	26,306	n/a	4,589	n/a	610
1957	n/a	31,221	n/a	5,093	n/a	596
1958	n/a	35,332	n/a	5,669	n/a	647
1959	n/a	38,134	n/a	6,615	n/a	714
1960	n/a	37,808	n/a	6,989	n/a	786
1961	n/a	35,860	n/a	7,977	n/a	943
1962	n/a	34,735	n/a	8,909	n/a	1,207
1963	n/a	33,458	n/a	9,460	n/a	1,378
1964	n/a	35,226	n/a	10,827	n/a	1,693
1965	n/a	36,691	n/a	12,246	n/a	2,124
1966	n/a	35,815	n/a	13,677	n/a	2,303
1967	n/a	36,186	n/a	13,887	n/a	2,614
1968	n/a	38,002	n/a	15,152	n/a	2,933
1969	n/a	39,972	n/a	14,980	n/a	3,387
1970	n/a	42,966	n/a	15,548	n/a	3,620
1971	1,565	43,167	2,930	16,383	741	3,640
1972	1,944	44,190	2,973	17,356	773	3,774
1973	2,136	43,429	2,551	17,152	708	3,587
1974	2,436	41,407	3,099	15,885	1,014	3,362
1975	2,468	38,210	3,250	15,773	891	3,138
1976	2,799	37,970	3,628	16,506	1,060	2,977
1977	2,996	40,095	3,825	16,551	995	2,813
1978	3,084	46,091	3,579	15,736	874	2,573
1979	3,788	52,598	3,944	15,624	929	2,815
1980	4,895	58,742	4,402	16,941	982	2,751
1981	5,622	62,935	4,589	17,643	1,054	2,841
1982	5,410	66,990	5,216	18,289	1,167	2,887
1983	6,151	72,471	5,145	19,673	1,179	3,023

NOTE: n/a = not available.
SOURCES: Data for 1950–1952 taken from *Facilities and Opportunities for Graduate Study in Engineering*, American Society for Engineering Education, Washington, D.C., March 1968. Data for 1953–1976 supplied by Engineering Manpower Commission, New York, N.Y. Data for 1977–1979 from *Engineering Manpower Bulletin #50*, Engineers Joint Council, New York, N.Y., November 1979. Data for 1980–1983 from Engineering Manpower Commission.

TABLE 2 Engineering Degrees by Field and Level, 1982–1983

Field of Engineering	Degree		
	Bachelor's	Master's	Doctor's
Aerospace	2,207	496	97
Agricultural	704	146	51
Architectural	568	30	0
Biomedical	577	178	50
Ceramic	294	55	18
Chemical	7,499	1,500	379
Civil	10,484	3,285	390
Computer	2,643	1,419	102[a]
Electrical, electronic	18,590	4,645	628
Engineering science	1,298	408	170
Environmental	292	456	68
Engineering, general	1,923	669	76
Industrial, manufacturing	3,808	1,400	108
Marine, naval architecture	698	144	18
Mechanical	16,484	2,964	399
Mining and mineral	1,019	280	54
Materials and metallurgical	1,085	561	228
Nuclear	420	301	114
Petroleum	1,420	225	14
Systems, operations	210	418	56
Other	247	93	3
Totals	72,471	19,673	3,023

[a]It should be noted that the number of graduates in "computer" fields is probably understated, for many computer science departments are not organizationally related to engineering schools and so may not be included in the reports as given by the Engineering Manpower Commission. The *Summary Report 1982: Doctorate Recipients From United States Universities* (ref. 15) lists 220 Ph.D. degrees in "Computer Science," as a subheading under "Physical Sciences." Since the latter report is based upon self-reporting of doctor's degree recipients, the number of 220 computer science Ph.D.s can probably be taken as a more accurate indication of the total number of computer-oriented Ph.D.s per year than the 102 shown in this table, some of which may be "Computer Science" and some of which may be "Computer Engineering." The problems that stem from self-reporting by individuals and from diverse organizational lines of reporting will cause the actual numbers to remain uncertain until definitions and organizational structures can be clarified.

SOURCE: Paul Doigan, "Engineering Degrees Granted, 1983," *Engineering Education*, April 1984, pp. 640–645.

At the beginning of World War II, the basic "science" of radar had long been understood, and working models existed, although they did not work as well as was desirable. It might be supposed that the development of more effective radar would be a classic sort of engineering task, since the basic science had been known for decades. Yet, at the Massachusetts Institute of Technology's (MIT's) Radiation Laboratory, where the major development work on radar was done, most of the 1,000 top-level participants were not engineers, but scientists temporarily working as engineers.[1] It had become apparent very early in the program that conventional engineering education programs of the day had not prepared most engineers to cope with the kinds of problems they faced with radar. For the most part, persons trained as physicists and mathematicians, usually at advanced-degree levels, did the job. Some observers believe that it was the research orientation derived from their educational backgrounds, as much as the additional exposure to science and mathematics, that enabled them to discover answers to problems no one had thought of before.[2]

Experiences such as this, coupled with a growing realization that the solution of a great number of postwar problems would depend increasingly on scientific knowledge, intensified the demand for inclusion of more science and mathematics in engineering curricula and stimulated a great expansion in graduate study and research. Before the establishment of the National Science Foundation in 1950, the expansion of both graduate study and research was spurred through initiatives of the armed services. These initiatives provided large-scale support for basic research in universities toward the end of and after World War II. This support for research laboratories (e.g., the Research Laboratory of Electronics at MIT[66]) also helped assure the availability of a pool of graduate engineers.

Grinter Report, 1952–1955

The move toward including more science in engineering was formalized by the publication in 1955 of the *Report on Evaluation of Engineering Education*, more familiarly known by the name of the chairman of the study committee as the "Grinter Report."[3] This report recommended strengthening of work in the sciences, strengthening of graduate programs, and development of superior engineering faculty members. The Grinter Report also recommended the following: integrated study of analysis, design, and engineering systems to enhance professional background; curricular flexibility; strengthening of humanities and social sciences in engineering programs; development of skills in speaking, writing, and graphic communication; and encouragement of experimental engineering. In terms of overall effect, however, the Grinter Report led to more science in engineering, stimulated the growth of graduate programs, and accelerated the trend toward more Ph.D.s on engineering faculties. The major growth trend in engineering graduate programs originated at about the time of the publication of this report.

PSAC Report, 1962

In 1962 the President's Science Advisory Committee published a report entitled *Meeting Manpower Needs in Science and Technology*.[4] The "PSAC Report" declared that the acceleration of graduate training in engineering, mathematics, and physical sciences, especially at the doctoral level, was a matter of urgent national priority requiring immediate action, without which severe shortages of engineers and scientists would occur. Engineering was identified as an especially crucial area. The federal government was to provide the funds needed, through increased research expenditures, provision of training grants, and fostering of new centers of scientific excellence. The country was, of course, reacting to shocks to its prestige caused by the success of *Sputnik*, and was also riding the crest of the greatest economic boom in its history, and these events simultaneously provided both the motive and the means for a major expansion in engineering graduate programs. Engineering education responded immediately, and the numbers of graduate students rose to unprecedented heights. (Just eight years later, the magnificent declarations of the PSAC Report were negated by a new conventional wisdom—that Ph.D.s were a drug on the market.)

Goals Study, 1963-1968

Close on the heels of the PSAC Report, the American Society for Engineering Education (ASEE) initiated the study, *Goals of Engineering Education*.[5] The "Goals Study" is probably the most ambitious, authoritative, and comprehensive study of engineering education ever undertaken. However, it suffered the misfortune of having been compiled during the very crest of the growth wave stimulated by the PSAC Report. As a result, it followed the prevailing philosophy of the time, used the latest data available (1966), and projected that the growth trends in engineering education would continue; it did so almost precisely at the time that the growth was in fact on the verge of being reversed.

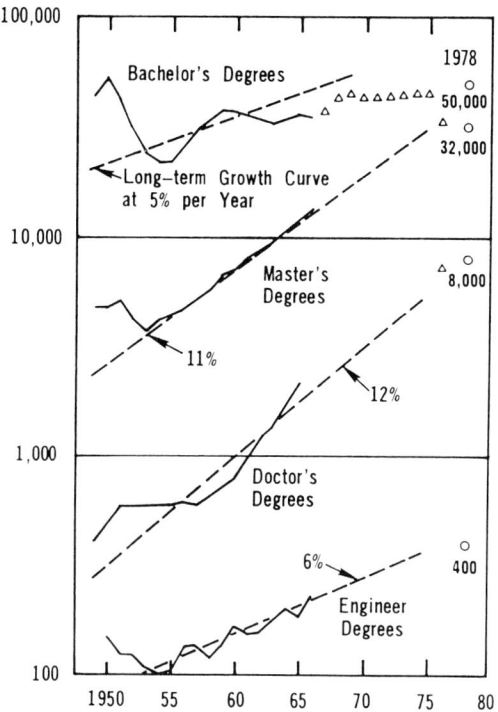

FIGURE 1 Engineering degrees in the United States, with projections. (Note: Triangles represent Office of Education projections; circles and accompanying numbers represent Goals Study predictions for 1978.) SOURCE: ASEE Goals Study (ref. 5).

The Goals Study staff observed that the number of master's degrees had been growing at a steady rate of 11 percent since 1950 and that doctor's degrees had been growing at the rate of 12 percent (Figure 1). These growth rates were extrapolated, with projections of 32,000 master's degrees and 8,000 doctor's degrees by 1978. In making these projections, the Goals staff was joined by the U.S. Office of Education which projected similar numbers. As it actually happened, however, the production of master's degrees slacked off after 1967 and went into a decline in 1973 (Figure 2). Subsequently, the growth rate has increased modestly, in concert with the enormous increase in undergraduate enrollment. Doctor's degrees peaked at 3,774 in 1972, and had recovered only to about 3,000 by 1983 (Figure 3). Projections for bachelor's degrees in the Goals Study were more modest than for graduate degrees, and are given here for completeness (Figures 1 and 4).

The Goals Study endorsed and reinforced the recommendations of

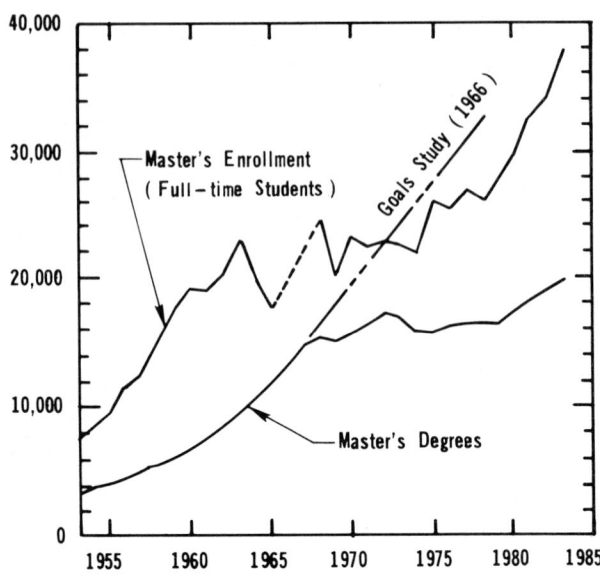

FIGURE 2 Engineering master's enrollment and degrees, U.S. totals, all schools. SOURCE: Data from Engineering Manpower Commission.

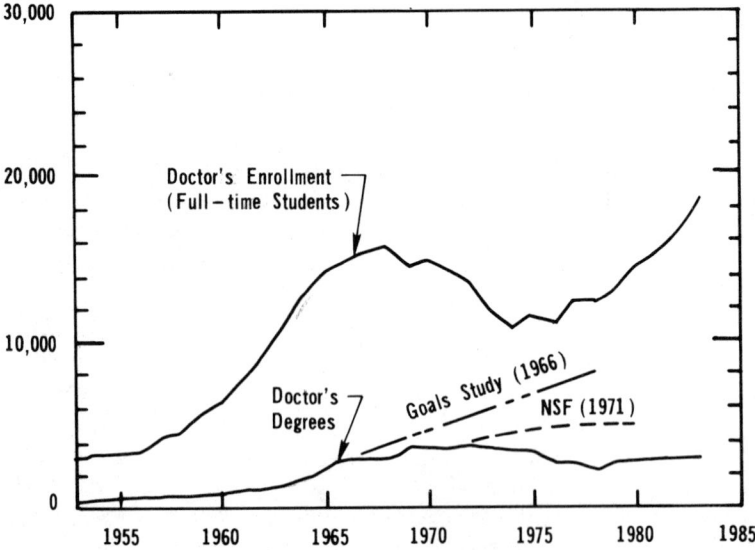

FIGURE 3 Engineering doctor's enrollment and degrees, U.S. totals, all schools. SOURCE: Data from Engineering Manpower Commission.

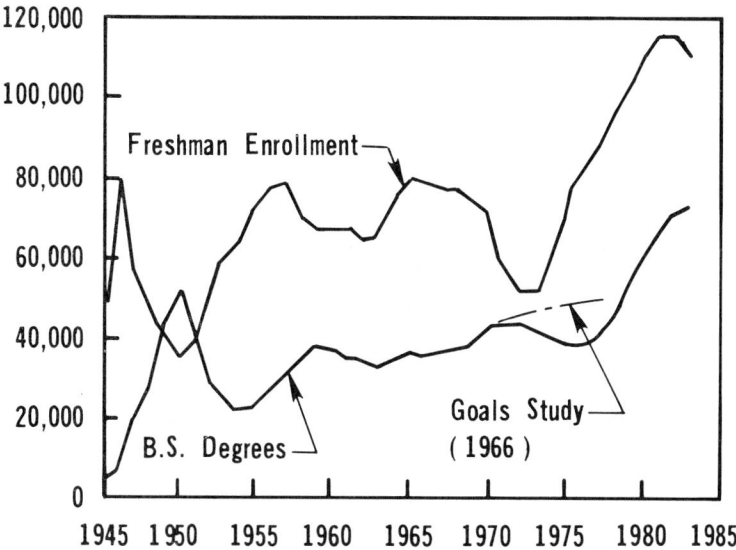

FIGURE 4 Engineering freshman enrollment and B.S. degrees, U.S. totals, all schools. SOURCE: Data from Engineering Manpower Commission.

the Grinter Report and added something new: that the master's degree should be more generally accepted as the basic degree in engineering. When this concept was first advanced, it was perceived as a proposal to replace the bachelor's degree with the master's degree, and was met by a storm of protest from both educators and employers. In response, the Goals Committee, in its final report, recommended that the importance of the bachelor's degree should be retained, but continued to insist that the importance of the master's degree should increase. In the years since publication of the Goals Report, the bachelor's degree has indeed maintained its importance, and it still represents the entry level for many kinds of professional engineering tasks. But the master's degree has also attained enhanced status as a degree of importance in engineering.

Retrenchment, 1969-1971

Several things happened simultaneously during the time identified here as the "retrenchment" period. First, large aerospace cutbacks occurred, creating unemployment problems for some segments of the engineering profession. Thousands of engineers and scientists, including many Ph.D.s, lost their jobs. Many of them remained unemployed for long periods of time.[6,7] Newspapers ran articles about unemployed engineers and scientists working as welders, rug salesmen, TV repairmen, bartenders, handymen, and operators of hot dog stands.[6,8,9] Second, newly graduated Ph.D.s in physics, chemistry, and mathematics began having difficulty getting jobs of their choice. Also, there were more new elementary and secondary school teachers than were needed, and many could not find jobs.[9] Third, the 1970 census showed that the number of young people in the college-age group was going to peak in the early 1980s and would decline thereafter, casting a pall on the prospects of those who were looking forward to university teaching careers. Fourth, the U.S. economy entered a period of economic recession, with the result that employers cut back on expenditures and postponed hiring new people. All of these events received a high level of media exposure, which produced an exaggerated and misleading picture of the employment picture for engineers.[10,11,12] The cumulative effects of these events on engineering education were drastic. Undergraduate enrollments plummeted (see Figure 4 for 1970–1973 period), but then recovered as it became apparent that the adverse publicity had been substantially misleading.

Other developments of the period were the termination of science development programs by the National Science Foundation (NSF), virtual cessation of training grants, and the emergence of an increasingly restrictive climate toward the funding of university research. The NSF published a report in 1969[13] stating that an oversupply of science and engineering doctorates by 1980 appeared unlikely; two years later NSF produced another report[14] reversing its earlier opinion, this time projecting that Ph.D. production by 1980 would result in an oversupply of 40 percent for engineers. The 1971 NSF projection is shown in Figure 3, along with the 1966 Goals Study projection. Both sets of projections far overshot the mark.

By 1980 not only had the oversupply of Ph.D. engineers projected by NSF failed to materialize, but a shortage had developed, at least from the viewpoint of academic institutions. Many engineering schools found it impossible to fully staff their faculties in the face of rapidly rising enrollments and the prevalent faculty environment. Now the nation must address the question: What is the appropriate relationship of engineering graduate study and research to the educational enterprise as a whole, and to the needs of industry, education, and government? Vital related questions are: How many engineering Ph.D.s should be produced each year? and Should programs be adopted that have the objective of increasing Ph.D. production?

3
Supply and Demand

The question of "demand" for engineers, whether undergraduate or graduate, has always been a difficult one. Many projections of demand are extrapolations from past trends. Some of these are quite sophisticated and include the effects of projected economic conditions and the movement toward high technology. Nevertheless, they are still extrapolations, and suffer from the basic problem of all extrapolations—they cannot anticipate surprises. Other projections are essentially opinion polls concerning the future and usually cannot retain their validity for more than a year or so.

The most comprehensive recent study of the scientist/engineer labor force was published by the National Science Foundation in 1984.[64] This study, which made projections based on various scenarios of economic growth, foresaw a general balance between supply and demand for engineers through 1987, with the exception of three fields: aeronautical/astronautical engineers, computer specialists, and electrical/electronic engineers. For these fields possible shortages were projected. Among the remaining engineering specialties, supply and demand for industrial and mechanical engineering were projected to be in rough balance, while all other engineering fields were projected to have personnel surpluses. All science, as distinguished from engineering, fields were projected to have surpluses. A difficulty for the present report is that the projections were not differentiated by degree level, so one cannot draw any information from them expressly regarding supply and demand for engineers with graduate degrees.

The present study cannot claim to be superior in accuracy to any other, and it deals with "demand" only to the degree that certain consequences could arise from current trends, which could affect the demand for Ph.D. engineers. However, it should be noted that Ph.D.s from some science fields have shown occupational mobility for activities in which engineering Ph.D.s are also sought, particularly in industry. This topic is developed below under the heading "Ph.D.s in Industry."

Supply of Ph.D.s

The "supply" of engineers is easier to quantify than is "demand," at least for four or five years into the future, because students presently enrolled in school can be counted. The number of engineering graduates with bachelor's degrees can be roughly predicted four years hence on the basis of freshman enrollments in the current year, although major changes in students' perceptions of future employment prospects can upset these predictions if the dropout rate changes significantly.

In the case of graduate degrees, one can make rough predictions of future master's degrees by assuming that the percentage of master's to bachelor's degrees awarded 1 year earlier will remain at about its recent level of 30 percent (see Figure 5). The ratio of doctor's to bachelor's degrees 5 years earlier* has ranged from about 6 percent to 10 percent in the last 15 years, and has remained in the range of 6 percent to 7 percent or so for the last 9 years. Thus, if students' perceptions of the attractiveness of the Ph.D. remain constant,[65] and in the absence of major programs to stimulate Ph.D. production, one could predict reasonably well the supply of doctorates 4 or 5 years into the future, based upon the percentage of bachelor's recipients in the recent past who subsequently earned Ph.D.s.

However, some evidence in the statistics shows that Ph.D. study is actually becoming less attractive to recent bachelor's graduates than was formerly the case. Table 1 shows that the number of engineering bachelor's degrees increased by 81 percent from 1977 to 1983. Table 3, on the other hand, shows that full-time doctoral enrollment increased only 47 percent in the same period.

---

*Reference 15 shows that the average *registered* time for engineering Ph.D.s in 1982 was 5.8 years from B.S. to Ph.D. Reference 16 shows that the median *total elapsed time* between the B.S. and Ph.D. ranged from 7.5 to 7.9 years during the period 1976–1981.

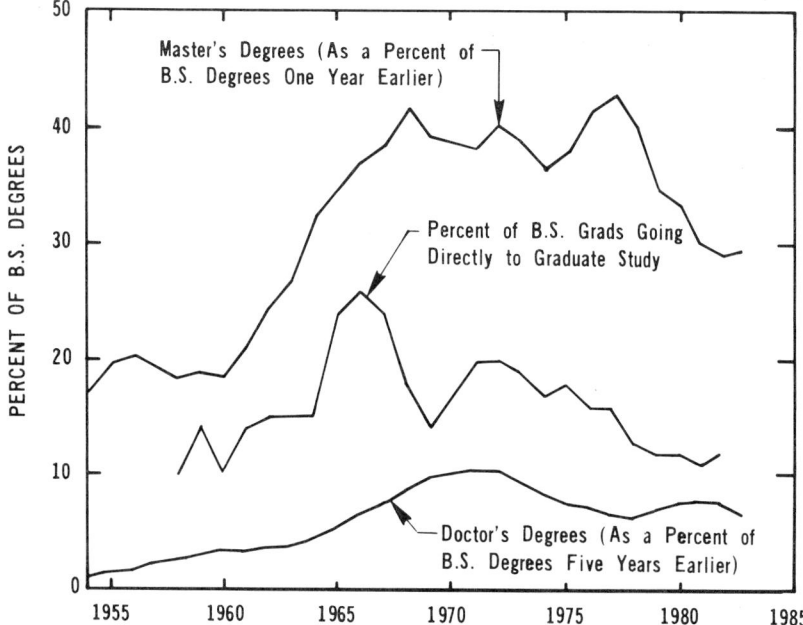

FIGURE 5 Ratios of master's and doctor's degrees to B.S. degrees. Source: Data from Engineering Manpower Commission.

Table 4 provides a way to estimate future Ph.D. production, at least until 1988, by examining the annual input of new doctoral students. The second and third columns of Table 4 ("total full-time doctoral enrollment" and "doctoral degrees granted") are taken from Table 3 and extended backward in time to 1967. The fourth column is the estimated number of continuing students. For example, the figure of 13,419 continuing doctoral students in the fall of 1983 is estimated by taking the total enrollment in fall 1982 (16,442 students) and subtracting from that the total number of doctoral degrees awarded during the academic year 1982–1983 (3,023 degrees). The difference between the total enrollment for fall 1983 (18,228 students) and the continuing students (13,419) must be the number of new doctoral students (4,809) who entered in the fall term of 1983. The estimated number of new students each fall is computed by this method and entered in the fifth column of Table 4. However, it can be seen that the behavior of these numbers is quite volatile: the first three figures in the fifth column are 3,325; 1,875; and 4,124, for example. Therefore, running three-year

TABLE 3 Full-time Doctoral Enrollment in Engineering, Number of Doctor's Degrees in Engineering, and Percent of Estimated Ph.D.s Available for Academic Employment, 1970–1988

	Full-time Doctoral Enrollment			Degrees			Estimated Ph.D.s Available for Academic Employment	
	Total	Foreign Nationals	Percent	Total	Foreign Nationals	Domestic	Percent[a]	Number
1970	14,802	—	—	3,620	—	—	31.3	1,133
1971	14,100	—	—	3,640	741	3,169	—	—
1972	13,460	—	—	3,774	773	3,001	34.6	1,306
1973	11,904	—	—	3,587	708	2,879	—	—
1974	10,628	—	—	3,362	1,014	2,348	29.5	992
1975	11,281	—	—	3,138	891	2,247	—	—
1976	10,963	—	—	2,977	1,060	1,917	36.2	1,078
1977	12,359	4,383	35.3	2,813	995	1,818	—	—
1978	12,321	4,273	34.6	2,573	874	1,699	36.0	926
1979	13,461	5,256	39.0	2,815	929	1,886	—	—
1980	14,465	5,995	41.4	2,751	982	1,769	34.9	960
1981	15,472	6,876	44.4	2,845	1,052	1,793	—	—
1982	16,442	6,756	41.1	2,887	1,167	1,720	32.8	947
1983	18,228	7,687	42.2	3,023	1,179	1,844	33.3	1,007
Estimated:								
1984	—	—	—	3,250	1,300	1,950	33.3	1,080
1985	—	—	—	3,400	1,360	2,040	33.3	1,130
1986	—	—	—	3,600	1,440	2,200	33.3	1,200
1987	—	—	—	3,750	1,500	2,250	33.3	1,250
1988	—	—	—	3,900	1,560	2,340	33.3	1,300

[a] *See* Table 6 of this report.
SOURCES: Enrollment data: *Engineering and Technology Enrollments* (New York: Engineering Manpower Commission, various years). Degree data: *Engineering and Technology Degrees* (New York: Engineering Manpower Commission, various years).

averages of the figures in column five have been computed and entered in the sixth column in order to smooth the data.

The figures from the sixth column ("estimated new students") of Table 4 have been plotted in Figure 6, together with the figures for annual engineering doctoral production. The shapes of the two curves are similar, both showing pronounced "troughs," but with the troughs displaced by five years. In Figure 7, the two curves of Figure 6 have been superimposed. A dashed line has been drawn through the "estimated new doctoral students" curve from 1980 on as an estimate of the manner in which the new student input is trending. A dashed line has been drawn parallel to that trend curve as an estimated five-year extrapolation of doctoral degree production. By this method, the engineering

SUPPLY AND DEMAND 21

TABLE 4 Growth in Full-time Doctoral Enrollment, 1967–1983

	Total Full-time Doctoral Enrollment	Doctoral Degrees Granted	Continuing Students (enrollment in year x, less degrees granted in year x + 1)	Estimated New Students (total enrollment less continuing students)	Estimated new students (running 3-year average)
1967	15,376	2,614	—	—	—
1968	15,768	2,933	12,433	3,325	—
1969	14,298	3,345	12,423	1,875	3,108
1970	14,802	3,620	10,678	4,124	2,979
1971	14,100	3,640	11,162	2,938	3,399
1972	13,460	3,774	10,326	3,134	2,701
1973	11,904	3,587	9,873	2,031	2,417
1974	10,628	3,362	8,542	2,086	2,636
1975	11,281	3,138	7,490	3,791	2,899
1976	10,963	2,977	8,143	2,820	3,606
1977	12,359	2,813	8,150	4,209	3,188
1978	12,321	2,573	9,786	2,535	3,566
1979	13,461	2,815	9,506	3,955	3,415
1980	14,465	2,751	10,710	3,755	3,854
1981	15,472	2,845	11,620	3,852	3,815
1982	16,442	2,887	12,585	3,837	4,166
1983	18,228	3,023	13,419	4,809	—

doctoral output of U.S. universities in 1988 is estimated to lie between 3,800 and 4,000.

Figure 6 shows an apparent anomaly in that the Ph.D. production in the years 1977, 1978, and 1979 appears to be larger than the number of new students who entered 5 years earlier. (This effect is most visible in Figure 7, where the curves have been superimposed.) There are two explanations for this apparent anomaly. One is that not all students take the same time to complete their degrees. The figure of a 5-year delay is probably a reasonable average, but some students will finish in 3 or 4 years after initial enrollment as doctoral students, while others may take 7 or 8 years. (The average of 5.8 years of registered time mentioned in the footnote on page 18 is from B.S. to Ph.D. and not from initial doctoral enrollment to Ph.D., which is the basis being used here.)

The second explanation of the apparent anomaly is more powerful than the first: the calculations in Table 4 only allow for the new students who enter in the fall term of each year. New doctoral students also enter during the winter and spring terms, and there is no way to enumerate these from the available data. It is assumed in this analysis that the effect of the midyear enrollees is a constant one, serving princi-

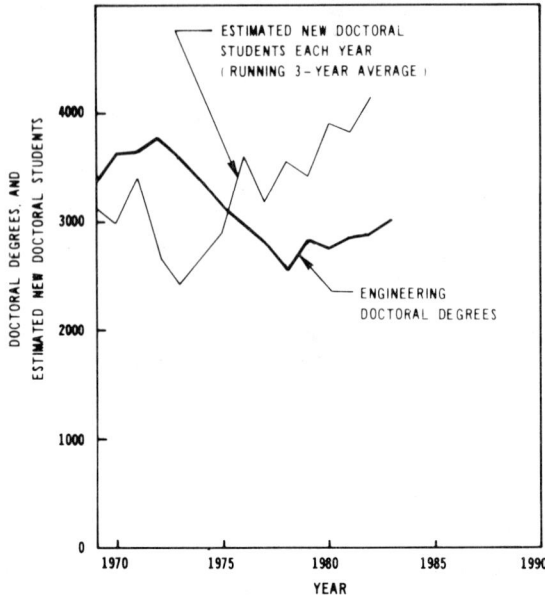

FIGURE 6 Engineering doctoral degrees per year and estimated new doctoral students per year (running 3-year average).

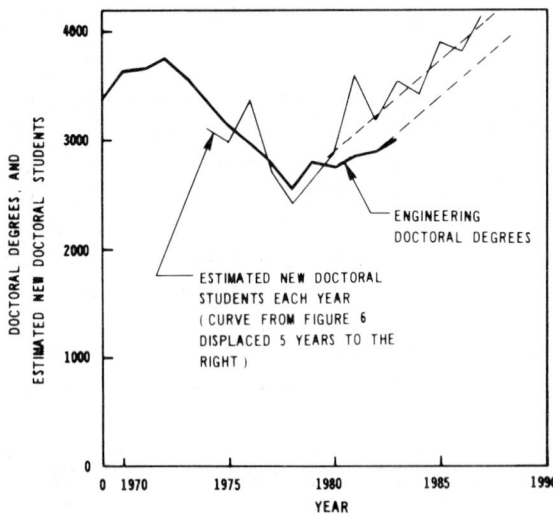

FIGURE 7 Engineering doctoral degrees per year, with curve for estimated new doctoral students per year from Figure 6 displaced five years to the right.

pally to shove the "estimated new doctoral students" curve in Figure 7 upward, but not materially changing the slope of the dashed lines upon which the estimates are based. (Note that the foregoing considerations do not apply to the "engineering doctoral degrees" figures, which include all the Ph.D. degrees granted during the year including those granted during the midyear period.)

The estimated figures for annual Ph.D. production from Figure 7 have been transferred to Table 3 (years 1984 through 1988). The proportion of foreign national and domestic Ph.D.s is also shown there using a constant 40 percent as the estimate for foreign nationals. Also, the numbers of Ph.D.s presumed to be available for academic employment are shown in the last column of Table 3, using a constant percentage of 33.3. The average number per year available for academic employment for the years 1984-1988 is calculated at 1,190 per year. The average number per year for the 1970s is calculated at 1,087 per year, from the entries in the last column of Table 3. Thus, even though the Ph.D. production of the country is rising markedly, the average annual number available for academic employment in the near future is only about 100 more than it was during the 1970s, when the engineering educational establishment was smaller than it is today. It was during the late 1970s and early 1980s that the shortage of engineering faculty developed, with the result that engineering schools reported 1,400 vacant faculty positions in 1982 nationwide.[23] The question is whether the future supply will be enough to meet the needs of educational institutions and industry simultaneously.*

Quality of Entering Graduate Students

The question of the trends in quality of engineering graduate students has frequently been raised. In particular, in view of the decline in the popularity of graduate study in the 1970s, some have wondered whether there was a corresponding decline in the ability levels of those who were admitted. One cannot form an absolute judgment on this matter, because the available national data only show what has happened to the ability level of the applicants as manifested by Graduate

---

* A survey taken in 1983 showed 1,570 vacant engineering faculty positions nationwide. *See* P. Doigan, "ASEE Survey of Engineering Faculty and Graduate Students, Fall 1983," *Engineering Education*, October 1984. The percentages of unfilled positions, by field, were as follows: Computer Science/Engineering—15.8 percent; Electrical Engineering—9.7 percent; Mechanical Engineering—7.7 percent; Chemical Engineering—7.1 percent; Civil Engineering—5.2 percent.

Record Exam (GRE) scores and not what happened to the selectivity of those admitted from the applicant pool. Also, the GRE scores apply to all those seeking engineering graduate study and do not permit us to separate those who terminate with a master's degree from those who go on to doctoral study.

GRE quantitative aptitude mean scores for prospective graduate students in engineering have ranged, during recent years, from a low of 649 (1974-1975) to a high of 665 (1972-1973).[18] The most recent available data show a quantitative mean score of 657 (1977-1978). The mean scores of prospective engineering graduate students have consistently been second only to students in the mathematical sciences, ranking just ahead of those in the physical sciences. In 1977-1978 the relative scores were as follows:

GRE (Quantitative)

Mathematical Sciences	669
Engineering	657
Physical Sciences	636
Life Sciences	559
Health Professions	517
Basic Social Sciences	514
Arts and Humanities	497
Applied Social Sciences	472
Education	449

In verbal aptitude GRE scores, engineers consistently rank near the bottom. In 1977-1978, the relative scores were as follows:

GRE (Verbal)

Arts and Humanities	532
Physical Sciences	517
Basic Social Sciences	516
Mathematical Sciences	504
Life Sciences	503
Health Professions	498
Applied Social Sciences	483
Engineering	459
Education	446

Engineering students, then, compete very well in the quantitative GRE, and there was little variation in their scores during the 1970s. They are much less competitive on the verbal GRE, although the data show a slightly improving trend in the 1970s.[19] (It is worth noting that the range of mean scores for the quantitative GRE is 220 points, whereas the mean verbal scores are more clustered, with a range of only

SUPPLY AND DEMAND 25

86 points.) The most that one can conclude from the data is that the quantitative ability of engineering graduate school applicants is high—among the highest—and is holding steady. The low scores on the GRE verbal scale may be a measure of the degree of difficulty and frustration many engineering students experience later with respect to career advancement. This situation cannot be expected to change unless steps are taken to improve students' communication skills, or unless engineering succeeds in drawing a larger share of high-scoring verbal applicants.

One small further insight regarding recent trends in the quality of engineering graduate students can be gained from an examination of the fractions of first- and second-decile bachelor's graduates at the University of Illinois (Urbana–Champaign) who go directly on to engineering graduate school. (The data include those who go to graduate school anywhere, not just those who continue at Illinois.) Figures for the most recent 12-year period are shown in Table 5.

The data in Table 5 for the first decile show that the peak was reached in 1973–1974, when 65 percent of the first-decile students went directly on to graduate school. Subsequently, this fraction declined each year until 1978–1979 and 1979–1980, when it reached a low of about 40 percent. Since then, the fraction has increased each year until

TABLE 5 B.S. Engineering Graduates at University of Illinois, Urbana–Champaign, Going Directly on to Any Engineering Graduate School, 1971–1972 to 1982–1983

	Number From:		Fraction From:	
	1st Decile	2nd Decile	1st Decile	2nd Decile
1971–1972	29	16	0.45	0.25
1972–1973	39	21	0.54	0.29
1973–1974	44	32	0.65	0.47
1974–1975	35	25	0.57	0.41
1975–1976	32	32	0.51	0.51
1976–1977	36	28	0.50	0.37
1977–1978	40	25	0.48	0.30
1978–1979[a]	26	19	0.39	0.29
1979–1980	42	40	0.40	0.38
1980–1981	57	39	0.48	0.33
1981–1982	67	55	0.54	0.44
1982–1983	79	46	0.64	0.37

[a]1978–1979 includes data on spring graduates only; fall data not available.
SOURCE: Dean's Office, College of Engineering, University of Illinois at Urbana–Champaign.

by 1982–1983 it was almost back to its former high, at 64 percent. The fraction in 1982–1983 from the second decile going directly to graduate school (37 percent) was the same as the overall average for the 12-year period. While no broad conclusions can be drawn from this information for the nation as a whole, Illinois is one of the country's major engineering schools, and the information provides some encouragement regarding the quality of students entering graduate school.

However, even though the Illinois figures are encouraging, they do not reveal how many of the brightest students are continuing on to doctoral study after completing a master's degree. Some engineering deans feel that not enough of the brightest students who are also U.S. citizens are pursuing doctoral study and that ways are needed to improve the attractiveness of doctoral work.

Ph.D.s in Academic Employment

The proportion of the nation's Ph.D.s that is available for academic employment can be estimated from Tables 6 and 7, which are taken from data collected by the National Research Council.[15] The Research Council periodically asks doctoral graduates about their postgraduation plans; the replies, by percentages of the total engineering doctorates for each year, are shown in Table 6, for selected years from 1960 to 1982. A 1981 follow-up study of doctoral graduates one year after graduation found that 100 percent of engineering doctorates seeking employment had been successful and that their actual type of employment closely matched their plans. Table 7, providing selected data from that follow-up survey, shows that 95.5 percent of those who had been seeking postdoctoral employment immediately after graduation were in academic employment one year later. Of those who were "seeking employment," 14 percent had gone into academic employment, 83.6 percent were employed in industry or government, and 2.4 percent were in "other" employment.

In Table 6, then, it is assumed that 100 percent of those who were planning to go into postdoctoral study would eventually enter academic employment and that 14 percent of those who were "seeking employment" would also enter academic employment. These proportions were used throughout the entire 1960–1982 span in Table 6 although they are actually known for only 1981, and then only for a sample of the total population. Based on this kind of estimation, the proportion entering academia in recent years is believed to be about one-third. The fraction of one-third has been employed in the "esti-

TABLE 6 Engineering Doctorates' Postgraduation Plans, 1960–1982 (percent)

	1960	1966	1970	1972	1974	1976	1978	1980	1982
Definite plans									
Academic employment	31.9	26.3	20.2	17.8	14.9	17.6	17.2	19.2	17.0
Industry and government employment	38.7	36.4	43.3	39.3	41.4	37.4	40.1	44.0	40.1
Postdoctoral study	3.5	5.4	5.2	8.6	6.7	10.2	10.5	9.1	7.6
Other employment	8.8	14.1	3.8	2.0	2.5	1.7	1.5	1.5	1.9
Seeking employment									
Postdoctoral study	0	0	2.8	5.2	5.1	5.3	5.8	4.4	5.3
Employment	15.5	12.6	21.9	21.7	20.1	22.1	18.0	15.4	20.6
Plans unknown	1.6	5.3	2.7	5.4	9.3	5.7	6.9	6.5	7.5
Estimated percentage in academic employment one year after receipt of doctorate									
Definite academic employment	31.9	26.3	20.2	17.8	14.9	17.6	17.2	19.2	17.0
Postdoctoral study (assume 100% enter academic employment)	3.5	5.4	5.2	8.6	6.7	10.2	10.5	9.1	7.6
Seeking postdoctoral (assume 100% enter academic employment)	0	0	2.8	5.2	5.1	5.3	5.8	4.4	5.3
Seeking employment (assume 14.0% enter academic employment)	2.2	1.8	3.1	3.0	2.8	3.1	2.5	2.2	2.9
Total % in academe	37.6	33.5	31.3	34.6	29.5	36.2	36.0	34.9	32.8

SOURCE: *Summary Report 1982: Doctorate Recipients From United States Universities* (Washington, D.C.: National Academy Press, 1983).

TABLE 7 Engineering Doctorates' Postgraduation Employment Plans in 1980, and Actual Status in 1981

Postgraduation Plans, 1980	Actual Status, 1981			
	Academic Employment	Industrial or Government Employment	Postdoctoral Study	Other
Postdoctorals	95.5%	0	4.5%	0
Seeking employment	14.0%	83.6%	0	2.4%

SOURCE: *Summary Report 1982: Doctorate Recipients From United States Universities* (Washington, D.C.: National Academy Press, 1983).

mated" portion of Table 3 to project an estimated supply of 1,080 to 1,300 doctoral engineers available annually for academic employment in the period 1984–1988.

Percentages of Ph.D.s estimated to be in academic employment for the years going back to 1970 have been transferred from Table 6 to the next-to-last column of Table 3 in order to compare the numbers presumed to be available in past years for academic employment with the numbers estimated for the future. As was mentioned previously, the average annual number projected for 1984–1988 is only about 100 greater than the average for the 1970s.

The data in Table 3 are those supplied by the Engineering Manpower Commission (EMC). However, the National Center for Education Statistics (NCES) and the National Research Council also collect data on doctor's degrees. Comparisons of these three sources are shown in Table 8. As can be seen, the figures from the three sources are mostly within 5 percent to 6 percent of each other, although in recent years they have deviated by 8 percent to 10 percent, with the EMC figures on the high side. The variations should not be surprising, since the three agencies collect the data using different methods. The EMC data come from deans' offices, the NCES figures are from registrars, and the Research Council figures are from reports filed by the individual degree recipients. Some of the differences in the figures probably arise from nonreporting, but there are also likely to be differences among the agencies in interpretations concerning the categories in which figures are to be counted. The field of computer science is a prominent example. Some universities lump those in computer science with engineers; some do not. For the purposes of this report, however, the figures are sufficiently close to one another so that the major conclusions are not affected. The data from the Engineering Manpower Commission have

TABLE 8 Engineering Doctorates, 1970-1983: Comparison of Numbers From Various Sources

	EMC	NCES	Ratio—NCES:EMC	NRC	Ratio—NRC:EMC
1970	3,620	3,691	1.02	3,434	0.95
1971	3,640	3,638	1.00	3,498	0.96
1972	3,774	3,671	0.97	3,503	0.93
1973	3,587	3,492	0.97	3,364	0.94
1974	3,362	3,312	0.99	3,147	0.94
1975	3,138	3,108	0.99	3,002	0.96
1976	2,977	2,821	0.95	2,834	0.95
1977	2,813	2,586	0.92	2,643	0.94
1978	2,573	2,440	0.95	2,423	0.94
1979	2,815	2,506	0.89	2,490	0.88
1980	2,751	2,507	0.91	2,479	0.90
1981	2,845	2,561	0.90	2,528	0.89
1982	2,887	2,636	0.91	2,644	0.92
1983	3,023	—	—	2,780	0.92

NOTE: EMC = Engineering Manpower Commission; NCES = National Center for Education Statistics; NRC = National Research Council.

SOURCES: *Engineering and Technology Degrees* (New York: Engineering Manpower Commission). *Projections of Education Statistics to 1988-1989* (Washington, D.C.: National Center for Education Statistics, 1980); plus computer run updates, 1979-1982. *Science and Engineering Doctorates: 1960-1981* (Washington, D.C.: National Science Foundation, NSF 83-309, 1983). *Summary Report 1982: Doctorate Recipients From United States Universities* (Washington, D.C.: National Academy Press, 1983).

been chosen for use here because the EMC reports degree data at all levels and also includes enrollment data.

An exception to the use of EMC data occurs in the case of foreign degree recipients. EMC reports these as "Foreign Nationals," but the Research Council figures, as reported by the National Science Foundation, place foreign students in two separate categories: (1) Non-U.S. Citizens, Permanent Residents; and (2) Non-U.S. Citizens, Temporary Residents.[16] Table 9 gives the Research Council figures, with the last column showing the ratio between the temporary residents and the total. This ratio had reached 42.1 percent by 1983. According to Research Council data, approximately half of the "temporary resident" engineering Ph.D.s plan to stay in this country after graduation. (The distribution of all engineering Ph.D.s in 1982, by citizenship, ethnic background, and sex, is shown in Table 10.)

The "needs" of educational institutions can be roughly estimated as

TABLE 9 Foreign Engineering Doctorates, 1970-1983 (National Research Council data)

	Total	U.S. Citizens	Non-U.S., Permanent	Non-U.S., Temporary	Unknown Citizenship	Percent— Temporary:Total
1970	3,434	2,514	430	471	19	13.7
1971	3,498	2,418	530	518	32	14.8
1972	3,503	2,330	622	519	32	14.8
1973	3,364	2,142	557	622	43	18.5
1974	3,147	1,752	515	704	176	22.4
1975	3,002	1,716	418	815	53	27.1
1976	2,834	1,557	390	813	74	28.7
1977	2,643	1,472	326	773	72	29.2
1978	2,423	1,261	325	768	69	31.7
1979	2,490	1,293	322	815	60	32.7
1980	2,479	1,255	299	851	74	34.3
1981	2,528	1,169	298	943	118	37.3
1982	2,644	1,165	296	1,028	155	38.9
1983	2,780	1,162	319	1,169	130	42.1

SOURCES: *Science and Engineering Doctorates: 1960-1981* (Washington, D.C.: National Science Foundation, NSF 83-309, 1983). *Summary Report 1982: Doctorate Recipients From United States Universities* (Washington, D.C.: National Academy Press, 1983). *Summary Report 1983: Doctorate Recipients From United States Universities* (Washington, D.C.: National Academy Press, 1984).

TABLE 10 Engineering Doctorates, 1982

	U.S. Citizens		Non-U.S. Citizens				Total (incl. those who did not report citizenship)	
			Permanent Residents		Temporary Residents			
	Male	Female	Male	Female	Male	Female	Male	Female
American Indian	3	—	—	—	—	—	3	—
Asian	69	3	168	6	545	20	806	29
Black	9	—	11	—	33	—	55	—
Hispanic	21	1	11	2	50	2	85	5
White	945	69	80	5	318	8	1,350	82
Other and unknown	44	1	12	1	51	1	221	8
	1,091	74	282	14	997	31	2,520	124

SOURCE: National Research Council, Survey of Earned Doctorates, Office of Scientific and Engineering Personnel, Doctorate Records File.

follows. Table 11 shows what happened between 1976 and 1982 to 51 large U.S. engineering schools* with respect to growth in enrollment and numbers of full-time faculty in the professorial ranks. During this period the number of students increased by 50 percent and the number of faculty by only 10 percent. Table 12 summarizes the data, showing that the "raw" student:faculty ratio for the aggregated 51 schools increased from 17.5 in 1975-1976 to 24.0 in 1981-1982, an increase of about 37 percent. This ratio is called "raw" because it results simply from adding the total full-time undergraduate and graduate enrollments and dividing the sum by the total number of faculty in the professorial ranks. A more useful measure of true workload would come from converting such numbers to so-called full-time equivalents, both for students and faculty. Since the formulas for such conversion vary from institution to institution and are not consistently available in a national data base, "raw" ratios are used here. Such raw ratios are useful, nevertheless, to show the major change that took place between 1976 and 1982.

The full-time enrollments in the 51 schools accounted for 50.4 percent of the total U.S. enrollment in 1975-1976, and 46.7 percent in 1981-1982. Using these percentages, the numbers of full-time faculty for all schools in the country were estimated by increasing the number 7,724 (1975-1976) by 1/0.504, and increasing 8,466 (1981-1982) by 1/0.467. This gave an estimated number of 15,320 faculty for 1975-1976 and 18,130 for 1981-1982. The latter number corresponds closely with the number of 18,000 faculty reported for all U.S. engineering schools in the fall 1981 survey of the Engineering College Faculty Shortage Project.[20] If the student:faculty ratio of 1975-1976 were to have been restored, approximately 6,750 additional faculty members would have been needed in 1981-1982. This has been labeled "estimated 1981-1982 faculty shortfall" in Table 12. An analysis by W. Edward Lear, using 1968-1969 and 1980-1981 as the comparison years, produced almost exactly the same number as the "shortfall": 6,700 faculty.[21]

Some have argued that there is no need to restore the 1975-1976 ratios and that engineering schools should find ways to increase their

---

* The 51 schools were selected as follows: first, all those schools with a full-time undergraduate enrollment greater than 1,300 in fall 1975 were placed on the list; next, any schools that were not already on the list but were in the top 10 Ph.D. producers in the years 1980-1983 (see Table 23) were added to the list; finally, since the sizes of faculty were taken from the publication *Engineering College Research and Graduate Study* for the years 1977 and 1983, any schools that were not listed in both years were removed from the list.

TABLE 11 Enrollments and Faculty for 51 Large U.S. Engineering Schools, 1975–1976 and 1981–1982

	1975–1976			1981–1982		
	Full-time Students		Full-time Faculty	Full-time Students		Full-time Faculty
School	UG	G		UG	G	
Arizona	1,762	334	93	3,308	434	105
Auburn	1,682	58	107	3,641	69	140
CCNY	2,218	79	71	2,893	258	77
Calif., Berkeley	2,241	1,538	201	2,150	1,577	227
Calif., Davis	1,360	269	82	1,535	613	104
Calif., UCLA	1,385	913	133	1,818	1,655	135
Calif. Polytechnic, San Luis Obispo	1,654	23	94	3,088	13	100
Cincinnati	2,182	359	108	2,277	453	98
Clarkson	1,718	101	56	2,192	145	70
Colorado	2,015	272	92	2,742	399	176
Colorado (Mines)	1,675	307	154	2,222	302	194
Connecticut	1,320	192	79	1,943	201	88
Cornell	2,269	721	218	2,383	872	202
Drexel	1,713	117	71	2,798	176	79
Florida	1,508	504	190	2,063	608	230
Georgia Tech.	4,051	470	263	6,667	698	253
Illinois, Chicago	2,080	200	76	2,912	495	80
Illinois, Urbana–Champaign	4,137	1,616	369	5,885	1,579	422
Iowa State	3,387	407	273	4,824	429	289
Louisiana State	1,807	173	101	3,652	309	127
Lowell	1,809	67	76	2,782	147	90
MIT	1,475	1,724	331	2,328	2,131	366
Maryland	2,018	195	111	3,424	228	121
Michigan	3,099	747	273	3,918	871	264
Michigan State	2,382	313	112	4,152	350	120
Michigan Tech.	2,471	84	125	4,683	124	127
Minnesota	2,342	500	188	4,072	737	200
Missouri, Columbia	1,420	284	106	2,342	167	129
Missouri, Rolla	2,780	258	182	4,930	153	184
N.C. State	3,087	227	145	4,659	311	174
N.J. Inst. Tech.	2,087	1,021	106	2,299	130	123
Nebraska	1,324	81	101	2,184	139	95
Northeastern	2,155	121	86	4,145	217	110
Northwestern	884	411	130	1,281	492	130
Ohio State	2,818	675	249	6,453	859	302
Oregon State	1,783	255	105	2,921	327	108
Penn State	4,866	426	307	6,613	520	253
Pittsburgh	1,590	176	100	2,041	291	100
Polytechnic Inst. of N.Y.	1,368	485	153	2,104	265	141

TABLE 11 (Continued)

School	1975-1976			1981-1982		
	Full-time Students		Full-time Faculty	Full-time Students		Full-time Faculty
	UG	G		UG	G	
Purdue	4,999	960	297	6,262	875	289
Rensselaer	2,225	405	122	2,751	777	161
SUNY, Buffalo	1,409	268	79	3,394	293	96
Stanford	783	1,313	141	1,279	1,686	158
Texas	3,041	808	177	5,678	874	169
Texas A&M	4,215	503	227	8,290	686	327
Texas Tech.	1,367	186	81	2,531	211	80
Univ. of Washington	2,287	496	180	2,610	688	183
Virginia Tech.	2,827	469	206	4,892	617	263
Washington State	1,362	191	137	2,935	295	117
Wisconsin	2,572	615	189	4,818	790	207
Worcester Polytechnic	1,503	77	71	1,968	141	83
Total	112,512	22,994	7,724	175,732	27,677	8,466

NOTE: UG = undergraduate; G = graduate.
SOURCES: Student enrollment: *Engineering and Technology Enrollments, Fall 1975* (New York: Engineering Manpower Commission, April 1976); *Engineering and Technology Enrollments, Fall 1981* (New York: Engineering Manpower Commission, 1982). Faculty: *Engineering College Research and Graduate Study, Engineering Education*, March 1977; *Engineering College Research and Graduate Study, Engineering Education*, March 1983.

productivity and to handle the increased student loads without requiring new faculty. The increased use of computers and of television is often cited as offering promising possibilities in this respect. (The section below on "Educational Technology and Productivity" discusses this matter in detail.) The reply of engineering educators generally is that computers and television are already being rapidly incorporated and utilized, but they point to the loss of quality that is occurring in much of the nation's engineering education delivery system primarily because of reduced student–faculty interaction.[22]

The increase in workload represented by the figures in Table 12 has had a number of consequences for engineering schools in addition to that of diminished student–faculty interaction. For some, the image of excessive overload has acted as a disincentive with regard to entering academic careers. In addition, there is a perception in some fields of engineering that opportunities for research are better in industry than in academia. This perception relates partly to the student overload and

TABLE 12 Comparison of Engineering Student:
Faculty Ratios, 1975–1976 and 1981–1982

	1975–1976	1981–1982
Student enrollment, 51 large schools		
Full-time undergraduate	112,512	175,732
Full-time graduate	22,994	27,677
Total	135,506	203,409
Full-time faculty (professorial ranks)	7,724	8,466
Raw student:faculty ratio	17.5	24.0
	291 schools	286 schools
Total enrollment		
Full-time undergraduate	231,379	387,577
Full-time graduate	37,285	47,772
Total	268,664	435,349
Percent of total, for 51 large schools	50.4	46.7
Estimated total faculty, all schools	15,320	18,130
Estimated faculty needed to restore 1975–1976 ratio	—	24,880
Estimated 1981–1982 faculty shortfall	—	6,750

partly to the inability of universities to provide sufficient funds to keep up with facilities needs, both for space and for equipment. These factors tend to aggravate the problems that keep universities from obtaining their "fair share" of Ph.D. production.

It has been argued that the current large enrollments in engineering and computer science constitute a "bubble" that will subside in the future. Past cycles in enrollments are cited in support of this view. And some have said that even if more faculty are needed, it is not necessary for them all to have Ph.D.s; this argument would tend to reduce the apparent need. Although these arguments have some validity, they will not solve the shortfall problem for the following reasons:

- National needs are such that engineering enrollments of the future should be maintained at a substantially higher level than was characteristic of the 1970s, even though enrollments may subside somewhat from the current high levels.
- Educators are strongly of the opinion that desirable student–faculty interaction has been interfered with as a result of high student:

faculty ratios. Maintenance of student–faculty interaction is vital to quality education, and more faculty will be needed even if student numbers decline from present levels.[22]

- A portion of the new faculty members needed by the country can function appropriately without Ph.D.s, but the large majority of faculty should be educated at the doctoral level. (This point is examined in greater detail in Chapter 8, "The Engineering Faculty.")

An additional factor affecting "need" is the anticipated near-term increase in the rate of faculty retirements. Engineering faculties expanded in the 1950s and 1960s, and many faculty members will be ready for retirement in the next 10 or 15 years. Table 13 shows the distribution of faculty in the professorial ranks by age, as provided by an "AAES/ASEE Survey."[20] Since the survey included 168 schools that responded, the distribution has been extrapolated on a pro rata basis to include the entire estimated population of 18,130 faculty for 286 schools, derived from Table 12. An estimated number of about 3,600 faculty are in the 56 to 65 age group, and most of them can be expected to retire in the next 10 years. There are approximately 5,400 in the 46 to 55 age group. These faculty members will begin retiring in substantial numbers about 10 years from now, and probably more than half will have retired in 15 years. Thus, it can be expected that about 7,000 faculty will retire in the next 15 years, an average of 450 or so per year, probably rising from 300 per year in the near future to 600 per year toward the end of the period.

Furthermore, there is the question of whether there is a net flow in, or out, between academia and industry. Data maintained by the Survey of Earned Doctorates (National Research Council) show the following:*

1981 Employer	1979 Employer	
	Business/Industry	University or 4-year College
Business/Industry	21,997	927
University or 4-year College	259	13,565

Thus, of those who were employed in industry in 1981, 21,997 had been in industry two years earlier, and 927 had been in academia. Of

---

* The numbers are based on a sampling survey, and the numbers given here are the *weighted* numbers, based on the distribution of the sample.

TABLE 13 Age Distribution of Engineering Faculty in Professorial Ranks

Age	Number of Faculty in Professorial Ranks (168 schools)	Distribution (%)	Estimated Number of Faculty in Professorial Ranks (286 schools)
70	31	0.3	54
66–70	239	2.1	381
56–65	2,196	20.1	3,644
46–55	3,287	30.0	5,439
36–45	3,315	30.4	5,512
25–35	1,861	17.0	3,082
25	12	0.1	18
Total	10,941	100.0	18,130

SOURCE: J. Geils, "The Faculty Shortage: A Review of the 1981 AAES/ASEE Survey," *Engineering Education*, November 1982, pp. 147–158.

those who were in academia in 1981, 13,565 had been in academia two years earlier, and 259 had been in industry. From this, we would conclude that the net flow had been outward from academia between 1979 and 1981, with the outflow rate 3.5 times as great as the inflow.

More recent data suggest that the flows have become approximately balanced. In 1981 the Engineering College Faculty Shortage Project reported that 266 faculty went from academia to industry and 251 went the other way, based upon responses from engineering deans.[20] In 1982, 227 went from academia to industry, and 252 went in the opposite direction.[23] In the present study, the flows are assumed to be in balance.*

There has been a great deal of discussion in recent years regarding the percentage of Ph.D. engineering students who are foreign nationals in the United States on temporary visas. In Table 9, it can be seen that this

---

*A survey taken in 1983 showed 300 going from academia to industry and 497 going from industry to academia. See P. Doigan, "ASEE Survey of Engineering Faculty and Graduate Students, Fall 1983," *Engineering Education*, October 1984. It would appear that the balance of flow has shifted in favor of academia since 1981, with a net flow to academia of 197. (Nevertheless, the same study reported that the total unfilled faculty positions had increased to 1,570, from the 1982 level of 1,400 unfilled positions.) Whether a net flow to academia can be regarded as permanent or whether it should be viewed as a transient response to crisis is open to question. Thus, for the purpose of the present analysis, the flows are assumed to be in balance.

SUPPLY AND DEMAND 37

percentage rose from about 14 percent in 1970 to 42 percent in 1983. During this same period, the absolute number of U.S. citizens in the doctoral pool dropped by more than half, from 2,500 per year to about 1,170. (The number of non-U.S. citizens on permanent visas has remained at about the level of 300 per year since 1977.) The Research Council surveys show that about half of all foreign engineering Ph.D.s expect to remain in this country, so they have constituted an increasingly significant portion of the engineering Ph.D. employment pool. Clearly, if these non-U.S. citizens had not been available, the engineering schools of this country would have faced an even more aggravated recruitment problem than was the actual case. A vital policy matter, however, is whether this country should be forced to rely on the importation of non-U.S. citizens in order to staff its engineering schools. It would appear that ways are needed to stimulate more U.S. citizens of high ability to enter the Ph.D. pool.

Ph.D.s in Industry

It is not possible to define the need for Ph.D.s in industry as clearly as it is to identify the need in academia. Nevertheless, in May 1984 an attempt was made to determine the general attitudes toward the Ph.D. by writing to top executives of 10 organizations known to be significant employers of engineers. All were asked if they perceived a trend toward increased recruitment of Ph.D. engineers and whether they had difficulty obtaining Ph.D.s of the quality needed; they were asked to provide typical starting salaries for Ph.D. and B.S. engineers.

Five of the 10 respondents stated that they did perceive an increasing trend in recruitment of Ph.D. engineers in their own organizations and in industry generally. The others responded that they saw no changes occurring in the demand for Ph.D.s, or that any trends they perceived were confined to certain special fields for which they had current need. Most respondents reported that they had no overall difficulty in recruiting Ph.D.s, although one did report having some difficulty, and another said there is a shortage of U.S. citizens among Ph.D. engineers. Several remarked that they were anxious to recruit people of the very highest quality, and two mentioned that they seek graduates from the top decile of the graduating cohort, either at the bachelor level or doctoral level.

The 1984 average monthly starting salaries reported by these respondents were as follows:

Type of Company	Ph.D.	B.S.
Chemical and petroleum	$3,284	$2,364
Computers, electrical manufacturing, aerospace, and others	3,271	2,240
Construction	2,750	2,175

These figures, in general, tend to be slightly higher than those reported nationally by the College Placement Council (CPC) in March 1984 (see Table 14).

It has been suggested that trends in starting salaries for new Ph.D.s, particularly as compared to starting salaries for new B.S. graduates, might give some insight into the demand for Ph.D.s by industrial

TABLE 14 Average Monthly Salary Offers to New Engineering Degree Candidates, 1965–1984

	B.S. Candidates				Doctoral Canadidates				Consumer Price Index (1967 = 100)
Year	Chem.	Civil	Elec. & Comp.	Mech.	Chem.	Civil	Elec. & Comp.	Mech.	
1965	$ 642	$ 618	$ 641	$ 635	$1,058	—	$1,166	$1,116	94.5
1966	682	658	679	670	1,102	—	1,203	1,136	97.2
1967	733	706	728	720	1,175	$1,128	1,261	1,212	100.0
1968	790	750	774	768	1,247	1,197	1,361	1,282	104.2
1969	849	797	826	820	1,319	1,234	1,379	1,309	109.8
1970	902	837	869	867	1,375	1,236	1,429	1,370	116.3
1971	920	850	877	881	1,395	1,102	1,388	1,278	121.3
1972	928	869	888	894	1,405	1,227	1,439	1,381	125.3
1973	962	908	931	927	1,438	1,298	1,508	1,418	133.0
1974	1,402	967	986	1,001	1,550	1,426	1,551	1,479	147.7
1975	1,196	1,064	1,081	1,122	1,645	1,382	1,550	1,624	161.2
1976	1,279	1,108	1,155	1,197	1,743	1,597	1,693	1,687	170.5
1977	1,389	1,185	1,245	1,286	1,882	1,625	1,811	1,777	181.5
1978	1,513	1,288	1,367	1,404	2,074	1,830	1,974	2,030	195.3
1979	1,642	1,402	1,520	1,536	2,231	2,260	2,221	2,160	217.7
1980	1,801	1,554	1,690	1,703	2,451	2,089	2,534	2,436	247.0
1981	2,030	1,775	1,882	1,908	2,745	2,523	2,793	2,659	272.3
1982	2,256	1,925	2,064	2,098	3,019	2,851	3,099	2,959	288.6
1983	2,228	1,869	2,128	2,096	3,130	2,936	3,200	3,150	299.3
1984 (March)	2,273	1,883	2,175	2,173	3,201	3,195	3,247	3,162	303.5

SOURCES: College Placement Council, Bethlehem, Pa. For Consumer Price Index: *Monthly Labor Review*, U.S. Dept. of Labor, March 1984, p. 82. Last value shown is for December 1983.

employers. Table 14 and Figures 8 through 16 provide information on this point.

Average monthly salary offers to new engineering degree candidates at the bachelor's and doctor's levels for 1965 through March 1984 are given in Table 14. The table only shows offers for chemical engineering, civil engineering, electrical and computer engineering, and mechanical engineering because these are the only engineering categories reported by CPC at the doctoral level. The average dollar offers for these four fields are plotted in Figures 8 through 11, where it can be seen that the dollar differentials between bachelor's- and doctor's-level offers in each field remained approximately constant until about 1977. At that time the differentials began to increase, with bachelor's-level offers increasing at about $140 to $150 per month per year (except for civil engineering, which increased about $115 per month per year), and doctoral offers increasing at about $210 to $220 per month per year. Figure 12 shows the differentials for all four fields from 1965 through 1984. In the years since 1980, civil engineering has increased its differential the most.

The effects of inflation tend to obscure the picture. Table 14 shows the behavior of the Consumer Price Index (CPI) for the years under

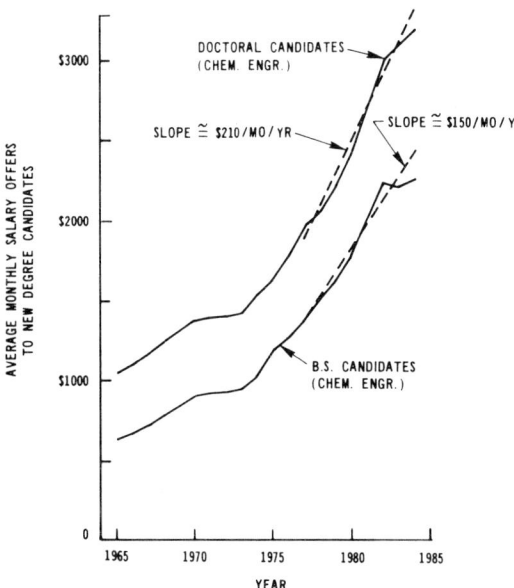

FIGURE 8 Average monthly salary offers to new chemical engineering degree candidates. SOURCE: Data from College Placement Council.

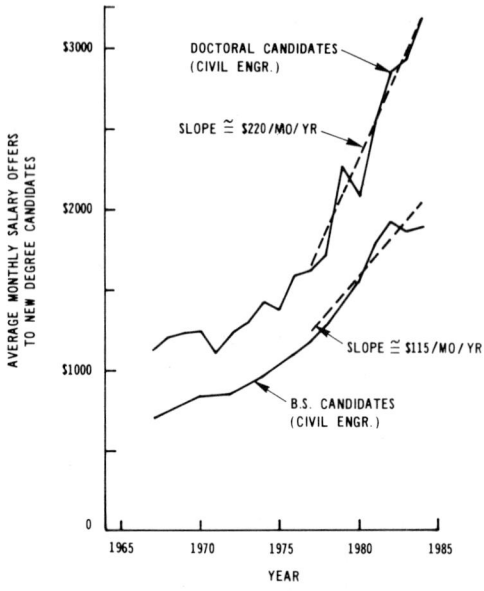

FIGURE 9 Average monthly salary offers to new civil engineering degree candidates. SOURCE: Data from College Placement Council.

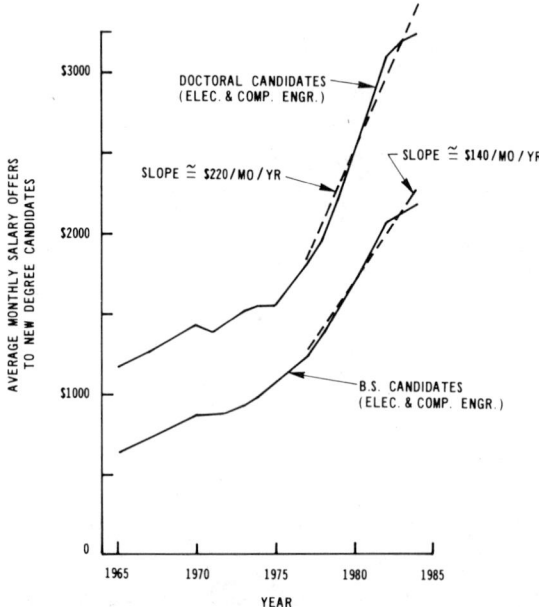

FIGURE 10 Average monthly salary offers to new electrical and computer engineering degree candidates. SOURCE: Data from College Placement Council.

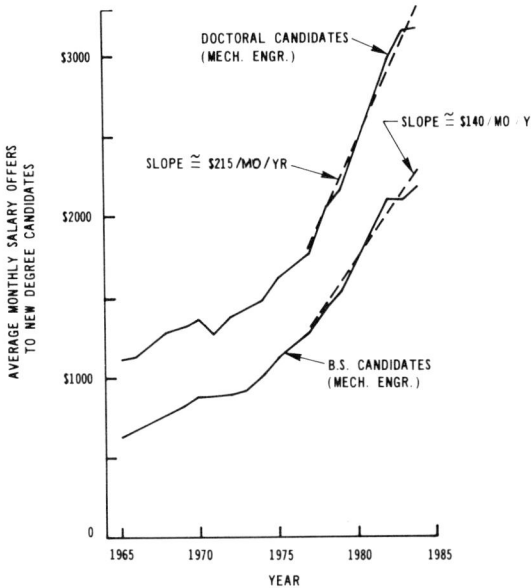

FIGURE 11 Average monthly salary offers to new mechanical engineering degree candidates. SOURCE: Data from College Placement Council.

consideration. All of the salary offers reported by CPC have been converted to constant 1967 dollars by use of the CPI, and replotted in Figures 13 through 16. These figures can be interpreted as the real value of engineering degrees to the market, particularly with regard to the relative market values of B.S. and Ph.D. degrees.

For all four engineering fields, there was an increase in the value of both bachelor's and doctor's degrees during the late 1960s, which was a period of intense recruiting for the space program. Subsequent years show a decline in value of the B.S. degree, followed by relative stability. A marked exception to this observation is civil engineering, which shows a gradual decline throughout almost the entire period.

The most pronounced effect to be observed in Figures 13 through 16 is the large decline in the relative value of the doctor's degree from the late 1960s until about 1975. The same figures show a gradual increase in the value of the doctor's degree since 1975, these trends being most apparent for civil engineering and for electrical and computer engineering. In the years since 1980 especially, an increase in value is apparent for the doctor's degree relative to the bachelor's for all four fields.

Not much can be concluded from these figures that is not already well known. The early 1970s, which was the period of marked decline

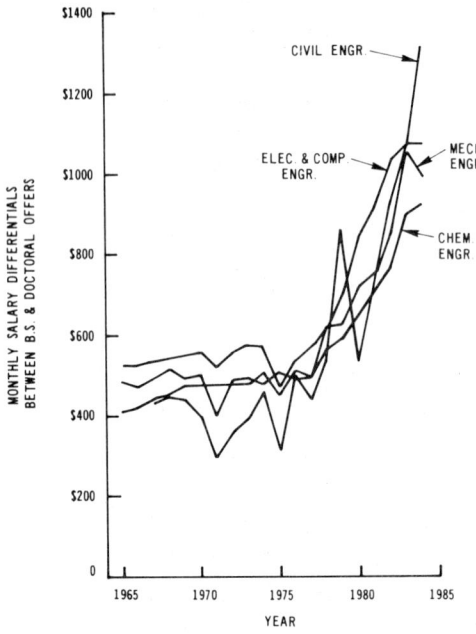

FIGURE 12 Comparison of average monthly salary offers to new B.S. and doctoral engineering degree candidates, for four engineering disciplines.

in the relative value of the doctorate, coincides with what was identified earlier as the "retrenchment" period; it was a period of relative unpopularity for engineering education generally. Many engineering schools struggled with the problems of low enrollment (see Figure 4), and a few engineering schools went out of business. It is not surprising that the "real" value of engineering starting salaries declined. The years since 1975 have seen surging enrollments and popularity for engineering. Perhaps the most surprising thing about Figures 13 through 16 is that they do not show a greater increase during recent years in the value of the B.S. degree in constant 1967 dollars. Given the publicity and recruitment activity for engineers, one would have expected a rising level of constant-dollar starting salaries. The fact that the doctoral offers do exhibit just such a rising trend in the last four years implies that recruiters are increasing their emphasis on doctoral degrees. However, the trends are too tenuous to draw any marked conclusions about scarcity.

Faculty salaries at many universities have been rising in recent years, but overall, salaries for assistant professors still lag significantly behind

SUPPLY AND DEMAND 43

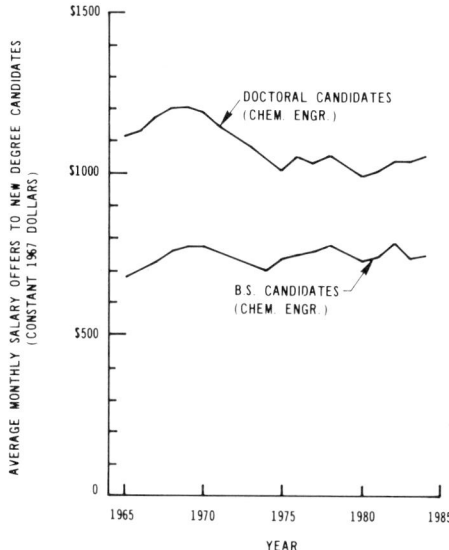

FIGURE 13 Average monthly salary offers to new chemical engineering degree candidates (data adjusted by Consumer Price Index to reflect constant 1967 dollars).

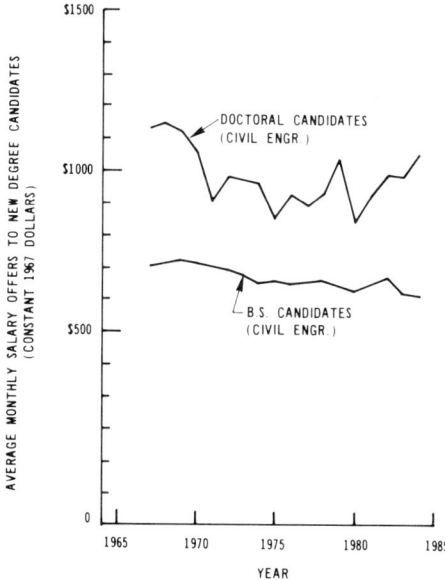

FIGURE 14 Average monthly salary offers to new civil engineering degree candidates (data adjusted by Consumer Price Index to reflect constant 1967 dollars).

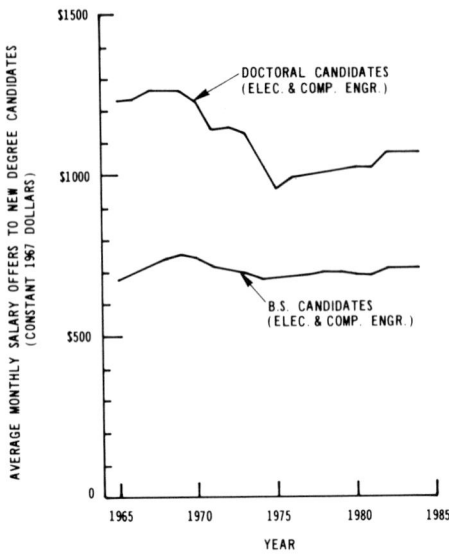

FIGURE 15 Average monthly salary offers to new electrical and computer engineering degree candidates (data adjusted by Consumer Price Index to reflect constant 1967 dollars).

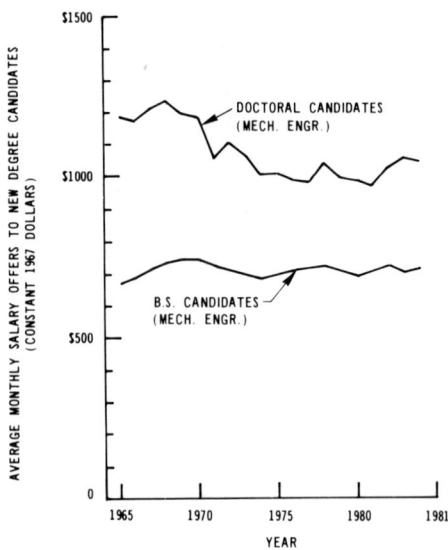

FIGURE 16 Average monthly salary offers to new mechanical engineering degree candidates (data adjusted by Consumer Price Index to reflect constant 1967 dollars).

SUPPLY AND DEMAND 45

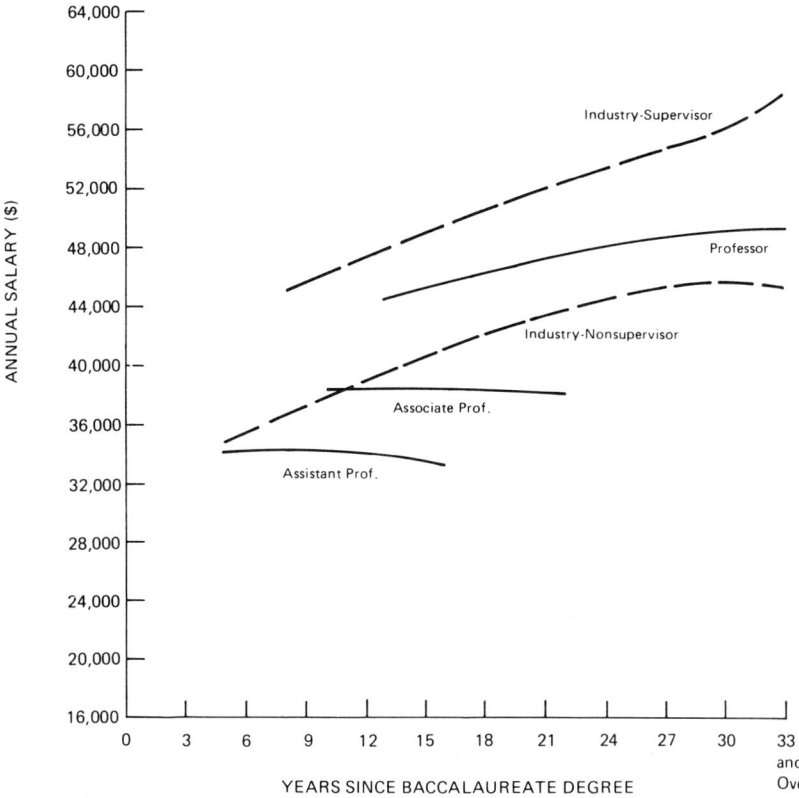

FIGURE 17 Comparison of academe–industry engineering Ph.D. salaries (all professorial salaries adjusted to 11-month basis). SOURCE: Data from Engineering Manpower Commission.

salaries for Ph.D. engineers in industry, even when adjusted to a full-year (11-month) basis. (See Figure 17.) In fact, salaries lag at all academic levels, since it may be argued that full professors should be compared with industry *supervisory* Ph.D.-holders and since some full professors are recruited into these positions. Figure 17 shows that the salaries of full professors are intermediate between those of nonsupervisors and supervisors in industry. Even these comparisons may be deceptive, however, because they involve median salaries. A crucial point is that for tenure-track positions schools typically attempt to hire the best doctoral engineers available. These same recruits can sometimes command significantly higher-than-median salaries in industry, so that the real disparity may be greater than the chart indicates.

It is worth noting that engineers by no means have a monopoly on the industrial need for doctoral-level personnel. The Research Council's Survey of Earned Doctorates shows that the number of Ph.D.s from the physical sciences that enter industry is about the same as it is for engineering.[15] For the 1982 graduates, 854 Ph.D.s from engineering had definite postgraduation plans for industrial employment, while 869 from the fields of chemistry, physics, and mathematics had similar plans (chemistry: 624; physics: 184; mathematics: 61).

The same report[15] compares the postgraduation plans of the 1980 graduates with their actual status one year later: Of those who had *definite* postgraduation employment plans in 1980 for going into industry, 97 percent were actually in industrial employment one year later, whether they were engineering graduates or physical sciences graduates. However, of those who were "seeking employment" in 1980, their actual status one year later was as follows:

Actual Status, 1981	Engineering Ph.D.s	Physical Sciences Ph.D.s
Academic employment	14.0%	7.8%
Industrial employment	82.9	44.3
Government employment	0.7	6.4
Other employment	2.4	4.6
Postdoctoral study	0.0	22.4
Not employed	0.0	14.5
	100.0	100.0

The conclusion to be drawn is that industry hires about equal proportions of Ph.D.s in engineering and in the physical sciences but that engineers appear to have a slight edge in attractiveness to industry; as shown above, 0 percent of the engineers seeking employment in 1980 were, one year later, in either the "not employed" or "postdoctoral study" category. Further, although one could not declare that there is a shortage of engineering Ph.D.s for industry, there does not appear to be a surplus either.

Increasing the Supply of Ph.D.s

It seems imperative for the nation to increase its supply of Ph.D. engineers. Table 3 and Figure 7 imply that this is indeed happening, with an increase from about 3,000 in 1983 to nearly 4,000 expected in 1988. However, as was mentioned, the number of U.S. citizens in this pool will remain substantially below the levels they represented in the 1970s unless actions are taken to encourage more of them to enter doctoral study.

Certain federal agencies have attempted to make projections of the future need for engineers. As mentioned above, the National Science Foundation published a report in 1983 that projected possible shortages in computer specialties, aeronautical/astronautical engineering, and electrical/electronic engineering, but the study did not assess the situation for persons with advanced degrees separately.[24,64]

In a separate study, also in 1983, the Department of Energy (DOE) assessed the adequacy of the supply of engineers and scientists for energy-related employment during the 1983-1988 period. The report concluded, relative to energy fields: "The potential for labor shortages during 1983-1988 is expected to be greatest at the Ph.D. level."[25] An increase of 20 percent in the supply of engineering Ph.D.s was projected by DOE, relative to 1982. Under the assumption that approximately half of the foreign nationals on temporary visas would remain in the United States, significant scarcities were projected for petroleum engineering and the earth/environmental sciences. However, under the alternate assumption that none of the students on temporary visas would remain in the United States, scarcities of 10 percent or more were projected for mathematics/computer science, chemistry, earth/environmental science, chemical engineering, nuclear engineering, petroleum engineering, mining engineering, and materials science.[25]

The matters of "shortage" and "supply and demand" are controversial. Many have pointed out that supply and demand necessarily become balanced at a price the market is willing to pay, so that one cannot properly speak of "shortages." On the other hand, engineering deans declare that the distress they have experienced in their inability to hire enough faculty for their needs is sufficient evidence of a "shortage." In reply, it has been said that academic employers need only improve salaries and working conditions sufficiently so that they can get their fair share of the existing Ph.D. production; if this were to occur, then an increase in total output would not be needed.

If we were concerned only with total numbers, the foregoing considerations would have considerable force. However, as has been seen, not enough of the brightest of U.S. citizens are motivated to enter the doctoral pool, and there are other ways in which the Ph.D. "market" deviates from an ideal one. The delay in the response from market stimulus to market response (five years or more for a Ph.D.) is enough in itself to interfere with an ideal response. An additional factor is that even if universities raise their salaries for engineering Ph.D.s to competitive levels—an outcome strongly recommended herein—the problem is not wholly solved, because the four or five years that must be spent during the period of graduate study with a marginal or submarginal income is enough to deter many new B.S. graduates from going on to graduate school. Improvement in graduate stipends is needed, at least to a level of 50 percent of what a new graduate could earn by going to industry instead of to graduate school. The level of 50 percent has become recognized by custom as a reasonable balance between the giving up of a salary of $26,000 or so (1984 level), and the opportunity to be paid to attend school full-time. (Some of the fellowships offered by federal agencies in 1984 were at the $13,000 to $14,000 level.)

There is a further disincentive for continuing on to graduate school that produces some concern. It is caused by the fact that many young people finish undergraduate school with large loan obligations and may wish to enter employment as soon as possible to begin reducing their debts. A solution to this disincentive might be to "forgive" such a loan if the individual goes on to complete a Ph.D.

While considering the disincentive of a financial burden carried by a student from undergraduate to graduate status, one should also consider the desirability of doctoral loans that are forgivable if the recipient enters academic employment for a specified number of years. There has been success in the past with forgivable loans of this nature, and such programs have been broadly supported because they focus financial aid on the location of greatest concern—academic employment. However, these arrangements might not be as attractive to students as might be supposed. From students' point of view, binding themselves to an obligation of academic employment several years in the future may not necessarily appear to be in their own best interest. No doubt some mix of forgivable loans and outright fellowships will prove to be most advantageous.

There is a question, too, in considering reliance upon natural market forces, about whether the country can afford to wait while the market works itself out to a condition of balance, especially in view of the present disincentives for attending graduate school. Many actions on many fronts are needed: universities must improve faculty salaries as well as their base of facilities, equipment, and support; industry needs to become involved in many ways, some of them financial; and a major fellowship program is needed to draw more of the top decile of B.S. graduates into doctoral study. One of the principal advantages of a fellowship program is that it shortens the time required to earn a doctorate, because students can attend school full time without needing to be employed in part-time jobs. Thus, the supply can be increased more quickly than by relying only on natural market forces. Fellowships can also be used to stimulate the entry of a greater portion of the top decile of U.S. citizens and permanent residents and thus lessen the country's dependence on importation of Ph.D. talent. If, for example, the nation wished 100 percent of its Ph.D. engineers to be U.S. citizens (or permanent residents), then an increase of 1,560 citizen Ph.D.s would be needed by 1988, according to the figures in Table 3. If, on the other hand, we were to return to the situation in 1972 when only 20 percent of the Ph.D.s were foreign nationals, then an increase of 700 to 800 citizen Ph.D.s would be needed in the projected output of 1988. (Since the current proportion of foreign Ph.D.s is 40 percent, and, since it is believed about half of the foreign Ph.D.s plan to remain in this country, a reduction of the foreign fraction to 20 percent would produce a condition of approximate balance with the fraction of foreign Ph.D.s who presently leave the country.)

In order to stimulate an increase of 700 to 800 more citizen Ph.D.s by 1988, new fellowships numbering substantially more than 700 or 800—perhaps 1,000— will be needed. This would allow for the attrition of those who do not complete their Ph.D. programs and also would make some allowance for those who would qualify for fellowships but who might have gone to doctoral study anyway, with or without a fellowship.

A fellowship program of this type undoubtedly will have a combination of federal and industrial support. Within the last two years, for example, the American Electronics Association (AEA) has established approximately 100 new Ph.D. fellowships, with a goal of 200. In approximately five years' time this program alone will increase Ph.D. production by 100 or so, but 100 new fellowships would have to be added each year for the next four years, until a total of 500 students is in the "pipeline," for the incremental output to be sustained in subsequent years at the level of 100 per year. In this program, a stipend of $10,000 per year plus tuition is offered to the fellow, and the student is expected to supplement this with employment as a research assistant or teaching assistant, plus at least one summer in industry. The AEA fellowships are of the "forgivable loan" type.

The AEA program is focused exclusively on electrical engineers and computer scientists. However, a recent informal survey by Hewlett-Packard produced the estimate that industry presently is offering approximately 200 to 300 new fellowship positions per year, mostly for engineering Ph.D.s, covering all disciplines. Most of these are of long standing and thus are part of the "constant" base. Perhaps only 100 or so are of recent origin.

The National Science Foundation provided the information that, of its 450 doctoral fellowship awards in 1982–1983, approximately 100 went to engineers. The NSF estimated that perhaps 150 new fellowships each year are offered by all other federal agencies combined, most of these going to engineers. The present stipend offered by NSF is $6,900 to the student, plus tuition and fees. The stipend was raised to $8,100 in the fall of 1984. Stipends being paid by some of the other federal agencies range as high as $12,500 to $14,500, with the higher stipends going to students in the third or fourth year of graduate study. Typically, in the federal fellowships, students are expected to engage in full-time study throughout the 12-month year.

The AEA fellowships are provided for four years. NSF offers three years of support, while the programs of the other federal agencies generally are based on support for four years. A few of the industrially sponsored fellowships cover up to five years of support.

If the fellowship stipend is 50 percent of starting salaries for B.S. engineers, then in 1984, for example, the stipend would have been $13,000 in the first year, and should be increased by a modest amount each year the student is in the program. With 1,000 new starts each year, and assuming a four-year program with some attrition each year, there might be 3,500 students actually in the program by the fourth year, when the program is fully under way. If the average stipend for all 3,500 students is $14,000 (1984 dollars), and if there is an accompanying grant to the institutions of up to $6,000 for tuition and fees, the cost per year would be in the range of $60 million to $70 million, divided between the federal government and industry.

A fellowship program provides only a part of the answer, giving an initial stimulus to prospective students and a "bridging" over the problem of financial support in graduate school. A permanent solution requires that universities take steps to make academic life more attractive than it has been recently by increasing salaries to competitive levels and reducing current overloads. Also, they need to provide modern laboratory space and equipment for both instructional and research purposes, so that students and faculty can have an opportunity to work with facilities that reasonably represent the state of the art.

The National Commission on Student Financial Assistance, in a 1983 report to the President and the Congress, recommended a substantial increase in the number of federally supported science and engineering fellowships. The commission also recommended substantial increases in funds for improving and modernizing university laboratories, equipment, and instrumentation.[61]

For its part, industry can help by continuing to increase its support of doctoral fellowships, giving state-of-the-art equipment, providing funding for laboratory renovations, increasing its support for university research, and providing grants to help support departmental expense such as that for computers, travel, and student projects. Federal and state governments can help industry in this regard by allowing generous tax deductions for contributions that help stimulate U.S. students to study graduate engineering.

Educational Technology and Productivity

It was mentioned earlier that developments in computers are significantly changing the manner in which engineering is practiced. Computer developments are also changing the way in which engineering education is practiced. Simultaneously, the instructional use of television has affected education with regard to both on-campus and off-campus use.

It is not the purpose of this report to review these and related developments in depth. That task has already been well carried out in the study *Educational Technology in Engineering*, prepared for the National Academy of Engineering by Lionel V. Baldwin and Kenneth S. Down.[26] However, our perception of need for additional faculty is heavily influenced by our vision of the prospects for improved educational productivity through the use of educational technology. If new technologies can somehow permit handling larger numbers of students with the current number of faculty and, of course, with no loss of quality, then an increase in engineering doctoral output would not be needed.

The simplest and most obvious measure of productivity is the student:faculty ratio, and the simplest way to increase this ratio is by grouping students in large classes, with the majority of student–faculty contact provided by teaching assistants (TAs). The technique is widely used by universities in handling large numbers of students in classes like physics, chemistry, and biology, but it is bitterly criticized by students (and their parents) because it deprives them of personal contact with the regular faculty. One of the earliest uses of instructional TV (ITV)—lectures by videotape, with the professors presumably thus made more available for personal contact—came essentially under the same criticism: the hundreds of students involved could not gain satisfactory contact with the professors in charge and so were shunted off to TAs anyway. Even though it has been shown that learning is not impaired in such courses, students exhibited strong objections to this kind of TV use and sometimes demonstrated their objections through declining attendance at the TV lectures.[26] However, in subsequent years more imaginative ways have been found to use ITV than merely producing the equivalent of large lecture classes. Probably the most effective of these has been in the form of supplementary instructional modules.

Videotaped supplementary instructional modules have been used in a variety of ways: for lecture review, making up missed lectures, classroom demonstrations, simulations, presentation of laboratory procedures, and self-paced instruction. The tapes are usually made available in individual study carrels at times that are convenient to the students. But even in this case students seemed to have demonstrated a preference for live tutors: in the NAE study cited above, one university found that only about 25 percent of its freshmen reported using supplemental lecture videotapes, although the system was wired into every dormitory room. Live tutors at the dormitories in the evening, on the other hand, drew an 80 percent response.[26] Nevertheless, ITV has been widely used for the purposes described above, in spite of heavy initial costs, principally for beginning-level courses in which the content is not subject to rapid obsolescence. However, in more advanced courses the need for regular revision of the material makes about as much demand on a professor's time as do the more conventional methods, thus offering no productivity gain.

The use of ITV for off-campus instruction brought something truly new to the educational scene, beginning at the University of Rhode Island in 1961. By 1980, 37 U.S. universities had adopted ITV, either "live" or by videotape, for engineering graduate study, for both credit and noncredit.

In the typical "live" TV mode, students at remote locations (usually at industrial sites) participate via TV in a class as it is being given simultaneously on campus. The cost-effectiveness comes about by saving time for off-campus students, who need not leave their places of employment to participate. However, there are extra administrative costs associated with the TV system: a camera operator must be hired; the TV system has maintenance costs; an operator must be in the system control room (as required by the Federal Communications Commission); provision usually is made for a "talk-back" system, probably through leased telephone lines; there must be a "courier" to carry homework and examinations back and forth; and additional office personnel are required to coordinate the system.

Instruction by videotape resembles that by simultaneous "live" telecast, with the exception, of course, that the students cannot ask questions of the professor during class. To compensate for this, proctors are usually provided by the industrial employer who receives the tapes.

Frequently the proctors are employees who took the course previously. If the courses change rapidly, which is the nature of graduate courses, such proctors can quickly lose touch with the course material. Nevertheless, some universities have approved videotaped courses for academic credit, prominent among them, Stanford University.[26]

Although the cost of maintenance of a TV transmitter is obviously avoided, there are special costs associated with videotape systems. Administrative costs are incurred in supervising and coordinating the systems, because tapes are constantly being sent to and received from many locations, generally by mail. Provision must be made, also, for sending and receiving homework and examinations. The delays associated with receiving such student work, correcting it, and sending it back have been the most troublesome aspect of using videotape systems. It is hoped that future reliance on transmission by satellite might alleviate such problems, but it has been estimated that satellite rental fees may range from $100 to $1,000 per hour, and satellite time must be provided for transmission not only of lecture material but also for student work if the present delays are to be overcome.

The recently organized National Technological University (NTU) began offering televised M.S. programs to a national audience in the fall of 1984. During the first year of operation, videotapes are being used, but it is expected that televised courses via satellite will be available in 1985. Long-term plans call for 80 graduate-level engineering courses to be offered per term, with approximately 9,000 students enrolled. The courses will originate from 18 member universities and will be distributed nationwide, but the degrees will be conferred by NTU. Courses can be received anywhere in the country once suitable "downlinks" to receive satellite signals have been installed at the receiving locations. Electronic mail and facsimile transmission will also be provided via satellite. A three–unit course will cost the student $1,000 to $1,400, with $600 to $1,000 of this going to the originating university and $75 to the instructor teaching the course.[27]

Computers are becoming ubiquitous in engineering primarily because they permit us to do things which were not possible before. It has become virtually impossible to design very large scale integrated circuits without computers, and structural analysis has been completely revolutionized by finite element analysis. New process plants are controlled by units that have computers at their hearts, and all of manufacturing is being revolutionized by robotics and computer-controlled methods. It is difficult to find any phase of engineering that is not being overturned by computer technology.

The advent of interactive computer graphics has had an especially strong and beneficial impact on engineering education. It has been pointed out, for example, that

> computer graphics is effective in engineering education because it emphasizes intuition rather than exact calculations. For years, engineering students used computers simply to get answers expressed to 10 decimal places. In order to understand the underlying relationships, the student generally had to print a number of solutions during one computer run and then try to interpret the tables of numbers on the crude plots from the line printer. . . . The pedagogical significance of changing with a light pen the location of a single charged particle in an electrical field with other charged particles, and watching *all* the field lines move as if they were rubber bands cannot be overstated! [Ref. 26]

Even as it becomes apparent that computers have deeply enriched engineering education, it also becomes clear that engineering education cannot rely entirely, or even extensively, upon prepackaged computer programs for educational purposes. Although it may be true that much of engineering in industry will utilize such programs, an educational curriculum relying excessively upon packaged programs will inculcate a "button-pushing" mentality on the part of the students and ill equip them to face new situations. Fundamental theory and mathematics must still be taught and learned, with computers interlaced to provide pedagogical improvements where appropriate.

Unfortunately, instead of lowering costs, computers have tended to increase them. The NAE study mentioned earlier concluded:

> Today, few people seriously consider lowering costs an argument for computing in instruction. The early literature abounds with cost-effectiveness discussions, but any honest comparison of computerized teaching costs with conventional teaching costs per hour are disappointing. . . . University-based advocates generally employ "anyhow" accounting—"we are going to do it anyhow"—when discussing costs. [Ref. 26]

One aspect of the cost of computers that has surprised and dismayed many engineering schools is that associated with technical support personnel and software maintenance. In the days when centralized computer centers represented the way business was done, the technical support personnel resided principally in the centers. But as minicomputers have proliferated, increased in power, and decreased in cost, computers are found everywhere, along with a bewildering variety of software systems. Individual academic departments are now finding that they need permanent support staff to manage these systems, particularly as networking enters the picture. Otherwise, the job falls on the shoulders of faculty members who are already overworked, compounding an already difficult situation.

Nevertheless, there *are* cases in which computers have aided instruction in a cost-effective sense, in a fashion analogous to that of videotape. Many beginning-level courses are taught by self-paced instruction; the student "contracts" to master certain modules of subject matter in a certain period of time. The student, going at his or her own pace, interacts with a computerized instructional module that provides pedagogical material selected in accordance with the student's rate of progress. Proctors are available to answer questions. When the material is supposedly mastered, the student takes a test from the proctor, which validates command of the material. Courses in calculus, statistics, elementary accounting, computer programming, and journalism have all been taught by such methods, or very similar ones.[26]

Findings and Recommendations

1. The nation can probably look forward to approximately 3,800 to 4,000 engineering Ph.D.s per year by 1988. Approximately 40 percent of these Ph.D.s are expected to be foreign nationals on temporary visas.

2. There has been little variation in the GRE scores of engineering graduate school applicants during the past decade. Engineering applicants consistently rank near the top in scores on the "quantitative" GRE, and consistently near the bottom in scores on the "verbal" GRE, when compared with applicants in other disciplines.

3. About one-third of new engineering Ph.D.s have entered academic employment in recent years. To maintain that fraction in future years, universities should take steps to make academic life more attractive than it has been recently for engineering faculty, in all ranks. The number of Ph.D.s available each year for academic employment during the next five years is expected to average only 100 or so more per year than was the case during the 1970s.

4. The percentage of non-U.S. citizens on temporary visas among engineering doctoral graduates has increased from 18.5 percent to 42.1 percent between 1973 and 1983. It is believed that about half of these graduates plan to stay in this country after graduation. In recent years, if there had been fewer foreign students in the employment pool, the difficulty for U.S. universities in obtaining engineering faculty would have been much more severe than it was. As a matter of national policy, it is questionable whether the United States should rely to such a degree upon the importation of Ph.D. talent.

5. The workload for U.S. engineering schools, as measured by student:faculty ratios, increased by about 37 percent between 1976 and 1982. To keep even with this growth would have required about 6,700 more faculty in 1981-1982 than actually existed (24,800 instead of the actual 18,100).

6. The image of excessive student overload has acted as a disincentive to some for entering academic careers. There is also a perception that opportunities for participating at the research frontier are diminishing in academic institutions, partly because of the student overload and partly because of the inability of universities to provide sufficient funds to keep up with facilities needs, both for space and equipment. These factors tend to aggravate the problems of universities in obtaining their "fair share" of Ph.D. production.

7. The view has sometimes been expressed that high engineering enrollments are a passing phenomenon and, in any event, that engineering schools could handle high enrollments by increasing their productivity and by hiring more non-Ph.D. faculty. The counterarguments are these:

   a. Enrollments of the future may subside somewhat from the current high levels but will be at a substantially higher level than was characteristic of the 1970s.
   b. Present student:faculty ratios are too high and interfere with student-faculty interaction; maintenance of a high level of such interaction is vital to a quality education.
   c. A portion of the new faculty members needed by the country can function appropriately without Ph.D.s, but the large majority of faculty should be educated at the doctoral level.

8. An estimated 3,600 engineering faculty are in the 56 to 65 age group, and an estimated 5,400 are in the 46 to 55 age group. Of these 9,000 faculty, perhaps 7,000 will retire in the next 15 years.

9. The flow of engineering faculty to industry is assumed to be approximately in balance with the flow in the opposite direction.

10. Industry employs Ph.D.s from many physical science fields as well as from engineering. However, engineering Ph.D.s seem to have a better chance for industrial employment. The data do not demonstrate that there is a shortage of engineering Ph.D.s for industry, but they do suggest that there is no surplus readily available for academia.

11. The supply of engineering Ph.D.s for academic employment is short enough that universities experience distress in faculty recruiting,

resulting in approximately 1,400 unfilled faculty positions in 1982 nationwide, and 1,570 unfilled positions for 1983.

12. In order to improve the faculty situation for engineering schools, several actions are necessary:

   a. The perception of academic life must be improved: universities must reduce the current high workloads, improve salary levels to competitive levels, and provide state-of-the-art facilities for instruction and research;
   b. The number of doctoral fellowships should be increased in order to increase the proportion of U.S. citizens from the top decile of their graduating classes who enter doctoral study. About 1,000 new "starts" should be available per year, with stipends at least equal to 50 percent of the average starting salaries of graduates going directly to industry. Industry and government should work together in providing this program. The total cost per year would be in the range of $60 million to $70 million for the nation.
   c. Industry, in addition to providing fellowships, should increase its financial support for engineering education, giving state-of-the-art equipment, providing funding for laboratory renovations, increasing its support for university research, and providing grants to help support departmental expense such as for computers, travel, and student projects. Federal and state governments can help by allowing generous tax deductions.

13. New developments in educational technology, principally involving computers and television, can be of major assistance in improving the quality and versatility of engineering education. Cost savings from such developments are not likely, however, and productivity improvements in the conventional sense of large student:faculty ratios have not so far materialized except at a cost to program quality.

4
Women and Minorities in Engineering

It is well known that women and ethnic minorities are underrepresented in engineering. Tables 15 through 17 show the following: women account for 13.2 percent of the bachelor's degrees in engineering, but only 4.7 percent of the doctor's degrees; blacks and Hispanics each account for about 2.6 percent of the bachelor's degrees, but blacks account for only 0.6 percent of the doctor's degrees, and Hispanics for only 1.4 percent. Asian/Pacific graduates, on the other hand, receive 4.3 percent of the bachelor's degrees and 5.7 percent of the doctor's degrees. Similar trends are visible in the production of master's degrees. Conclusions cannot be drawn from the small numbers of Native American students.

The falloff between bachelor's and doctor's degrees for women, blacks, and Hispanics may be explained in large part by the intense recruiting pressure to which these groups are subjected upon graduation with the B.S. Another part of the explanation is that the "pipeline" is still being filled.

Table 17 shows the production of bachelor's degrees among women and minorities between 1978 and 1982, and thus suggests trends in the graduate school "pipeline supply." For every group the trend is up, both in absolute numbers and percentages of the total. The increase is especially marked for women, although they are still far from their representation in society as a whole. The percentages for blacks and Hispanics leave these groups far underrepresented. It is difficult to draw national conclusions for the Asian/Pacific group because 50 percent of

TABLE 15 Engineering Doctor's Degrees—Women and Minorities, 1978-1983

	1978	1979	1980	1981	1982	1983	1983 (%)
Total doctor's degrees	2,573	2,815	2,751	2,841	2,887	3,023	—
Women	51	61	88	90	126	142	4.7
Blacks	15	19	19	16	11	19	0.6
Hispanic	25	22	25	20	26	41	1.4
Asian/Pacific	175	177	154	148	124	173	5.7
Native American	3	—	1	3	2	0	0.0

SOURCES: *Engineering and Technology Degrees* (New York: Engineering Manpower Commission, 1978, 1979, 1980, 1981, 1982). Paul Doigan, "Engineering Degrees Granted, 1983," *Engineering Education*, April 1984, pp. 640-645.

TABLE 16 Engineering Master's Degrees—Women and Minorities, 1978-1983

	1978	1979	1980	1981	1982	1983	1983 (%)
Total master's degrees	15,736	15,624	16,941	17,643	18,289	19,673	
Women	794	866	1,083	1,225	1,539	1,782	9.1
Blacks	199	157	163	182	184	258	1.3
Hispanic	239	214	246	276	215	306	1.6
Asian/Pacific	784	675	807	959	836	1,283	6.5
Native American	4	9	4	7	15	16	<0.1

SOURCES: *Engineering and Technology Degrees* (New York: Engineering Manpower Commission, 1978, 1979, 1980, 1981, 1982). Paul Doigan, "Engineering Degrees Granted, 1983," *Engineering Education*, April 1984, pp. 640-645.

TABLE 17 Engineering Bachelor's Degrees—Women and Minorities, 1978-1983

	1978	1979	1980	1981	1982	1983	1978 (%)	1983 (%)
Total bachelor's degrees	46,091	52,598	58,742	62,935	66,990	72,471		
Women	3,280	4,716	5,680	6,557	8,140	9,566	7.1	13.2
Blacks	894	1,076	1,320	1,445	1,644	1,862	1.9	2.6
Hispanic	1,072	1,245	1,332	1,513	1,608	1,883	2.3	2.6
Asian/Pacific	1,195	1,532	1,922	2,267	2,577	3,098	2.6	4.3
Native American	37	59	60	90	91	97	<0.1	0.1

SOURCES: *Engineering and Technology Degrees* (New York: Engineering Manpower Commission, 1978, 1979, 1980, 1981, 1982). Paul Doigan, "Engineering Degrees Granted, 1983," *Engineering Education*, April 1984, pp. 640-645.

these students are concentrated in just three states: California (32 percent), New York (13 percent), and Hawaii (6 percent).[28]

One can look back even farther in the pipeline and examine what has happened since 1978 in engineering freshman enrollments. Table 18 gives the figures. Again the trends are up, and again, notably so for women. The absolute number of women freshman students dropped in 1983, but then so did the overall number of freshmen. The net result was that the percentage of women freshman students increased slightly in 1983 to 17.0 percent, from the 1982 level of 16.6 percent. A troublesome factor is the falloff in 1982 for blacks and Hispanics, following several successive years of increases. A possible explanation is the condition of the economy in those years, which may have had an adverse impact upon the financial ability of persons in those groups to attend college. Also, many schools were limiting undergraduate enrollment, which may have had an adverse impact on minority applicants.

The ability of the system to produce representative numbers of women and ethnic minorities at the graduate level is dependent upon how many students from these groups enter engineering at the freshman level. Studies have shown that the pool from which future Ph.D.s in science or engineering will come has essentially been established by the conclusion of high school.[29] In fact, from one-third to one-half of the future science-oriented students had selected science as their field of interest by as early as ninth grade. Therefore, it is clear that efforts to increase the representation of women and minorities in engineering must be concentrated in high school or earlier. The factors are both cultural and economic. Members of minority groups especially are likely to be disadvantaged economically, so the provision of financial

TABLE 18 Engineering Freshman Enrollments—Women and Minorities, 1978–1983

	1978	1979	1980	1981	1982	1983	1978 (%)	1983 (%)
Total freshman enrollments	95,805	103,124	110,149	115,280	115,303	109,638		
Women	11,789	14,031	16,004	18,238	19,155	18,689	12.3	17.0
Blacks	5,493	6,339	6,661	7,015	6,715	6,342	5.7	5.8
Hispanic	3,296	3,768	3,208	4,768	4,421	4,760	3.4	4.3
Asian/Pacific	2,169	3,133	2,889	4,035	4,098	4,983	2.3	4.5
Native American	225	317	365	412	371	376	0.2	0.3

SOURCES: *Engineering and Technology Enrollments* (New York: Engineering Manpower Commission, 1978, 1979, 1980, 1981, 1982). Paul Doigan, "Engineering Degrees Granted, 1983," *Engineering Education*, April 1984, pp. 640–645.

TABLE 19 Persistence Rates of Engineering Students

	Freshman Enrollments Fall, 1979	B.S. Degrees 1982–83	Ratio: "Persistence Rate"
All students	103,724	72,471	0.70
Women	14,031	9,566	0.68
Blacks	6,339	1,862	0.29
Hispanic	3,768	1,883	0.50
Asian/Pacific	3,133	3,098	0.99
Native American	317	97	0.31

support at all levels of education is vital. In addition, minorities are more likely than other groups to be disadvantaged educationally, so programs aimed at this problem are also vital. There is a tendency for minority high school students to be unaware of engineering as a possible career even though engineering is the second largest profession after teaching. It has often been said that engineering is the "invisible profession" as far as high school students are concerned. Thus, high school and junior high programs should be expanded, especially at inner-city schools, to expose students to the variety of opportunities in engineering and to encourage them to take the courses in mathematics and science that are necessary for entering engineering school. These choices must usually be made at the sophomore and junior levels in high school, and may simply not be possible later. Thus, the principal efforts at "filling the pipeline" should be made at the high school level. A number of programs in various states focus on these objectives. A more concerted national effort is needed, under the guidance of the National Action Council for Minorities in Engineering (NACME).

Another factor influencing the availability of women and ethnic minority students for graduate school admission is their "persistence rate" in undergraduate school.[38] Table 19 provides some data on this, using 1979 and 1983 as the comparison years. Table 19 shows a "persistence rate" of 0.70 for all students, computed by taking the number of B.S. degrees nationally in 1982–1983, divided by the number of freshmen four years earlier. There are two things wrong with such a simple ratio. First, not all students graduate in four years; second, the "freshman enrollments" figure does not include community college enrollments, so the computed ratio taken by itself is much too high. However, for rough comparisons between groups, these ratios are acceptable. The persistence rate for women is seen to be close to that for all students. But, for all minority groups except Asian/Pacific, the persistence rates are markedly lower than for all students. (The ratio of 0.99 for Asian/Pacific students is probably misleading, since community college figures are not included in the freshman enrollments. For example, 32 percent of the Asian/Pacific students are in California, which has a very large community college system.) It has been learned from experience that many minority students have academic difficulties in engineering school as a consequence of both educational and cultural disadvantages. Often, their inner-city schools have not prepared them well in mathematics and science. Then, as academic trouble begins to develop, they frequently do not know where to turn for help. For undergraduate students, and especially for freshmen, universities generally look like massive, unresponsive monoliths, even though many counseling and teaching resources may in fact be available. Thus, for minority students especially, active intervention by counselors who monitor progress and give sympathetic assistance is vital. Engineering schools need to develop "retention programs" for their minority students that will provide sympathetic, personal counseling for students to help them get started at the right places in their course work and to guide them to learning-assistance services offered by their campuses. Financial aid should be made available to economically disadvantaged students at all levels of engineering education. Since minority students frequently come from backgrounds of economic disadvantage, such aid programs are vital to making it possible for minority students to gain access to college and to remain there.

Some evidence that, for women at least, the pipeline is being filled is given by the graphs of Figure 18. The percentage of women at every educational level increased between 1978 and 1983. If the curves continue to rise, the percentage of women in engineering will reach overall representational proportions eventually, but it is impossible to predict at this point whether that will actually happen. It does appear, however, that young women no longer perceive engineering as a field that is closed to them.

Women in Academic Careers

The pool of faculty women who are engineers is too small to have been investigated statistically in any of the studies that could be identified as appropriate references for this report. For this reason, much of the material that follows is derivative in nature. For example, when a

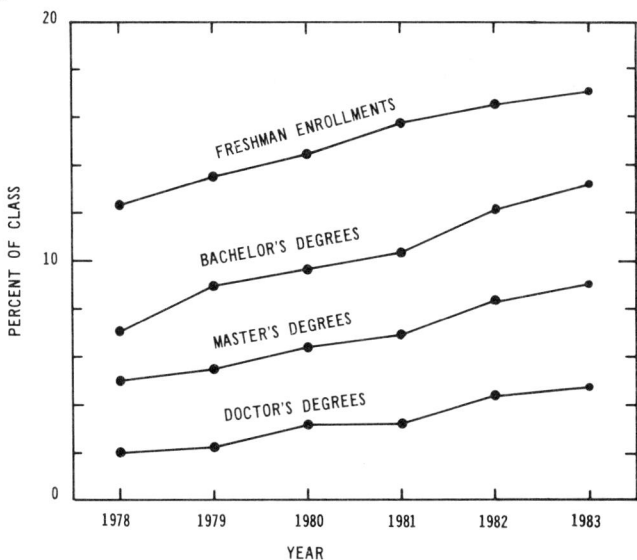

FIGURE 18 Women engineers as a percentage of year's class. SOURCE: Data from Engineering Manpower Commission.

body of data concerns women in science and engineering, it is assumed that faculty women are included unless otherwise specified; when studies about academic women in science reveal features that appear to be discriminatory toward women, it is assumed that these features also describe the situation of academic women engineers; when a salary survey carried out for engineers in all areas of the workplace reveals generally lower salaries for female members of the profession, faculty women are presumed to be similarly affected; and when a body of sociological studies documents discriminatory forces that influence the situation of women in society in general, those same forces are presumed to influence the careers of faculty women in colleges of engineering. Valuable material concerning perceptions and experiences can be found in the *Proceedings of the Engineering Educators Workshop*—the workshop was held in conjunction with the 1979 national convention of the Society of Women Engineers[30]—and in the proceedings of a 1975 conference held at Cornell University, *Women in Engineering—Beyond Recruitment*.[31]

Women make up a very small percentage of the total number of engineering faculty members in this country—less than 0.9 percent in 1979, according to the Scientific Manpower Commission. The talent pool from which faculty are hired is closely related to the number obtaining their doctorate in engineering. Table 15 shows a small but steady increase in the numbers of such women. It seems reasonable to suppose that the numbers will continue to increase over the years immediately ahead because of the increase in bachelor's degrees. However, it cannot be automatically assumed that future faculty positions accepted by women will be directly proportional to these increases.

In the fall of 1980, 49.4 percent of all undergraduates in U.S. institutions of higher education were women, and women comprised 42 percent of all graduate students. In the same academic term, 12.8 percent of all engineering undergraduate students were women, as were 8.8 percent of all engineering graduate students. The relatively low rate of continuation of women into engineering graduate school, compared with that of women in other disciplines, cannot be explained in terms of a simple lack of interest in graduate studies on the part of the women. The attractive job offers available to engineering B.S. graduates must certainly be a factor, but other factors also appear to be at work.[32]

The 1980 reported figures on financial support of students in engineering graduate school show the following:[32]

Percent With Financial Support

	Total Number	Institutional Support	Federal Support	Self-Support	Unknown
Men	38,399	30.0	27.0	28.4	14.6
Women	4,052	31.0	21.6	34.0	13.4

Women compare favorably with men in gaining support from their own institutions, but receive a substantially smaller proportion of federal support. While the disparity cannot, of course, be proved to be attributable to discrimination, it also cannot automatically be assumed that these figures prove women less capable than men. Suggestive evidence that discrimination may indeed be a factor comes from a 1980 study of a profession other than engineering, but similar in employing mostly men. In this study, conducted by the Academy of Management Review, it was shown that male applicants for scholarships were judged "more intelligent" and "more likeable" than their female counterparts, and it was shown that male applicants for a study-abroad program had been favored over female applicants.[33,34] A series of studies by Rosen and Jardee[35,36,37] yielded similar results. Male applicants in several contexts were more likely to be regarded as acceptable for hire, evaluated more favorably with regard to potential, judged more suitable for promotion, more likely to be chosen to attend a professional training conference, and more likely to have their recommendations accepted regarding resolution of a supervisor–subordinate conflict. Moreover, the sex differences were larger with respect to evaluations of candidates for demanding jobs than for undemanding jobs.

Barriers to the full participation of women in merit awards and prize programs appear to fall into two categories: attitudinal barriers on the part of nominators, judges, and sometimes the women themselves; and procedural barriers such as letters of recommendation, interviewing techniques, and criteria statements, which, often inadvertently, exclude women disproportionately as competitors or winners, and exclude women as nominators or judges.[33] Because of the perception of women as a group in society, women often face greater difficulty than men do in matters of fair evaluation, especially when the evaluators are male and are concerned with predicting future performance. Such might be the case when a male faculty member must decide which of several potential research students shows promise worthy of support on his grant, or when a male administrator decides which members of a pool of applicants for teaching assistantships are most likely to handle these responsibilities well.[39] Male faculty are often likely to perceive women students as less capable and less professionally committed than men.[33]

The Scientific Manpower Commission has reported that there were 11,868 faculty members in engineering colleges in 1979, of which 102 were women.[32] Of these, 100 women were in tenured or tenure-track positions. (See table below.)

Engineering Faculty Members in U.S. Universities and Colleges

	Professor	Assoc. Professor	Ass't. Professor	Other
Men	6,162 (52.4%)	3,661 (31.1%)	1,644 (14.0%)	294 (2.5%)
Women	15 (14.7%)	27 (26.5%)	58 (56.9%)	2 (1.9%)

SOURCE: Scientific Manpower Commission.

On the other hand, the National Research Council reported 17,100 engineering faculty members in 1979, of whom 200 were women. (See Table 20.) Since the total number of engineering faculty members in the United States is known from other sources to be approximately

TABLE 20 Tenure Status of Academically Employed Doctoral Scientists and Engineers by Field of Ph.D. and Sex, 1979

1979 Tenure Status	All Fields	Field of Doctorate										
		Math.	Com-puter	Physics/ Astron.	Chem.	Environ.	Eng.	Agric.	Med.	Biol.	Psych.	Soc. Sci.
Total academically employed	166,900	12,400	800	13,700	15,600	4,700	17,100	7,900	4,900	37,000	18,200	34,600
Tenured	59.2%	71.2%	30.8%	53.4%	60%	62.7%	63.9%	68.1%	48.7%	51.3%	54.1%	65.4%
Tenure track–not tenured	16.4	15.1	42.9	11.5	11.1	11.6	14.4	14.4	21.9	17.0	18.6	19.9
Nontenure track	13.3	6.7	7.8	16.9	16.3	14.1	10.9	9.1	16.1	18.7	15.3	7.9
No report	11.1	7.0	18.5	18.2	12.6	11.6	10.7	8.4	13.3	13.0	12.0	6.8
Years to tenure	6.1	5.9	4.8	7.2	7.1	6.4	5.3	5.5	6.7	7.0	6.4	5.0
Total academically employed, male	146,700	11,500	700	13,300	14,200	4,500	16,900	7,700	4,100	30,700	13,800	29,300
Tenured	62.5%	73.1%	32.4%	54.0%	62.5%	64.4%	64.2%	69.2%	51.4%	56.2%	60.1%	69.3%
Tenure track–not tenured	15.2	14.4	42.6	11.5	11.0	11.6	14.3	14.1	21.3	16.0	15.7	17.9
Nontenure track	11.7	5.9	6.4	16.5	14.6	12.7	10.9	8.5	15.0	15.6	13.0	6.6
No report	10.5	6.6	18.6	18.0	11.8	11.3	10.6	8.2	12.3	12.3	11.2	6.1
Years to tenure	6.1	5.9	4.9	7.2	7.08	6.3	5.3	5.5	6.8	6.9	6.4	5.0
Total academically employed, female	20,200	900	100	400	1,400	200	200	200	800	6,300	4,400	5,300
Tenured	35.2%	47.4%	13.9%	32.9%	35.0%	28.1%	29.8%	31.3%	35.2%	27.3%	35.1%	43.7%
Tenure track–not tenured	24.9	23.9	45.8	10.4	11.5	11.3	28.0	24.2	24.8	21.9	28.0	31.0
Nontenure track	24.8	16.4	22.2	31.9	33.1	43.4	17.3	27.3	21.6	33.5	22.5	15.0
No report	15.1	12.3	18.1	24.8	20.4	17.2	25.0	17.2	18.3	17.3	14.4	10.3
Years to tenure	6.3	6.0		9.5	8.0	7.7	6.0	4.3	5.9	7.8	6.7	4.8

SOURCE: *Science, Engineering, and Humanities Doctorates in the United States, 1979 Profile* (Washington, D.C.: National Academy of Sciences/National Research Council, 1980).

18,000,[23] it is assumed the latter source is closer to the truth. In either case, the number of women faculty is small—on the order of only 1 percent.

Data in the table ("Engineering Faculty Members in U.S. Universities and Colleges") from the Scientific Manpower Commission show markedly different distributions in the academic ranks for men and women. To some degree this is due to the more recent entry of women faculty into engineering education, but another study, from the National Science Foundation, suggests that this is not the whole explanation. NSF reported in 1982 that women with doctorates in scientific and engineering fields were less likely than men to be tenured or in tenure-track positions (59 percent as opposed to 78 percent), and that of the tenured faculty, 53 percent of the women and 75 percent of the men held the rank of associate or full professor. These differences were found to persist after adjustments were made for field, year of receipt of doctorate, and quality of the institution from which the doctorate was granted.[40]

According to the same NSF study, a larger proportion of the women than of the men in the combined doctoral scientist and engineer pool are employed by educational institutions. Within universities, women were found to be less likely than men to be in research institutions. A study by the National Academy of Sciences found that almost 26 percent of doctoral men but only 21 percent of the doctoral women scientists were in the top 50 academic institutions, as measured by R&D expenditures.[41] A larger proportion of the women than of the men were found in two-year colleges and in elementary and secondary schools. There is no obvious way to separate these combined scientist and engineer figures to isolate the case of women engineers, except to note that women engineers would more likely be found in industry than in secondary and elementary schools. However, there is also no evident reason to suppose that the status of women engineers in academia is different from that of women scientists.

Table 20 provides some comparative data on time to tenure for men and women in various academic pursuits, including engineering. Of the 200 women in the engineering category, 57.8 percent are listed as tenured or on the tenure track. Their years-to-tenure figure is 6.0; that for men is 5.3. Physics, astronomy, and chemistry show even greater differences, as do biology and environmental studies. Women appear to obtain tenure sooner than men do in agriculture, medicine, and social science. In the humanities fields (Table 21), the average time to tenure for women is slightly less than for men—5.0 years, compared to 5.2. Thus, there are some academic fields, including engineering, in which

TABLE 21 Tenure Status of Ph.D.s in the Humanities by Field of Ph.D. and Sex, 1979

		Field of Doctorate									
1979 Tenure Status	All Fields	Hist.	Art Hist.	Music	Speech/ Theater	Phil.	Other Human.	Eng./ Amer. Lang. & Lit.	Class. Lang. & Lit.	Modern Lang. & Lit.
Total academically employed	53,200	12,900	1,300	3,400	3,200	4,400	1,800	15,400	1,200	9,600
Tenured	67.9%	73.4%	59.2%	60.8%	72.3%	63.1%	60.2%	69.0%	64.3%	64.6%
Tenure track–not tenured	14.7	10.2	22.5	17.6	15.5	15.8	17.5	14.7	17.5	17.2
Nontenure track	11.2	12.0	10.5	15.0	6.0	12.3	13.9	10.3	10.5	10.8
No report	6.2	4.3	7.7	6.6	5.9	8.8	8.3	5.0	7.2	7.5
Years to tenure	5.2	5.3	4.7	4.2	5.4	5.1	5.0	5.1	6.3	5.2
Total academically employed, male	40,600	11,000	700	2,800	2,700	3,900	1,300	10,900	900	6,400
Tenured	73.8%	77.4%	69.4%	63.8%	75.1%	66.6%	64.3%	77.2%	73.6%	72.8%
Tenure track–not tenured	12.8	9.4	19.0	17.6	14.4	15.0	13.3	12.0	14.8	14.9
Nontenure track	8.4	9.2	5.3	13.3	4.9	10.5	13.2	7.0	6.3	7.0
No report	5.0	4.1	6.3	5.2	5.6	7.9	9.2	3.8	5.2	5.3
Years to tenure	5.2	5.3	4.4	4.2	5.5	5.0	5.1	5.2	6.1	5.3
Total academically employed, female	12,600	1,900	600	600	500	500	500	4,500	300	3,200
Tenured	48.7%	50.5%	45.6%	45.2%	58.6%	38.5%	50.6%	49.2%	37.9%	48.7%
Tenure track–not tenured	20.9	15.1	27.3	17.2	20.6	21.9	27.6	21.3	25.6	21.7
Nontenure track	20.3	28.5	17.5	23.7	13.3	24.9	15.6	18.9	23.2	18.0
No report	10.1	5.8	9.6	13.8	7.5	14.8	6.2	10.6	13.3	11.6
Years to tenure	5.0	5.2	5.6	4.2	4.6	5.7	4.8	4.8	7.1	5.1

SOURCE: *Science, Engineering, and Humanities Doctorates in the United States, 1979 Profile* (Washington, D.C.: National Academy of Sciences/National Research Council, 1980).

women typically wait longer for tenure than men do. Given the comparative data about other fields, it does not seem reasonable to attribute this delay to the commitments of women to their homes, children, or personal lives or to any other factor intrinsic to women. In fact, a study reported in 1981 revealed that not only do marriage and family life fail to hamper the scientific productivity of academic women in pure science, they appear related to increases in the research performance on which academic advancement is typically based. "Women scientists who are married turn out to be significantly more prolific than those who are not; and women who are married with either one or two children are slightly more scientifically productive than unmarried women, and only slightly less so than those who are married without children."[42]

Salary studies related to the salaries of 250,000 members of the Institute of Electrical and Electronics Engineers (IEEE), including electrical engineering faculty women, generally reveal lower salary structures for women, as do data related to the salaries of software workers in industry.[42,43,44] Figure 19 gives the data from the IEEE and shows annual pay differentials ranging from $3,700 to $8,300. After the IEEE data were adjusted to account for such possible modifying features as later entry of women into the field, level of education and professional responsibility, the differential remained at $2,600. In the case of software workers, the authors of the survey concluded that it is a $5,000-a-year liability to be a woman in this field. (See Figure 20.)

In spite of their low representation as engineering faculty members (approximately 1 percent), women submitted 5 percent of the research proposals received by the Engineering Directorate of the National Sci-

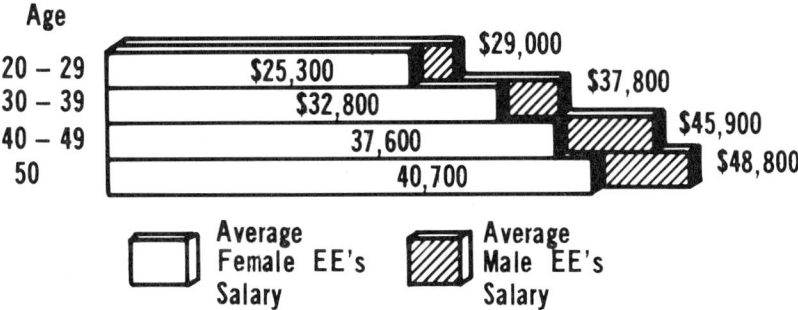

FIGURE 19 Average yearly pay for male and female electrical engineers, according to age. SOURCE: *The Institute*, Vol. 8, No. 3, March 1984. © 1984 IEEE. Reprinted with permission.

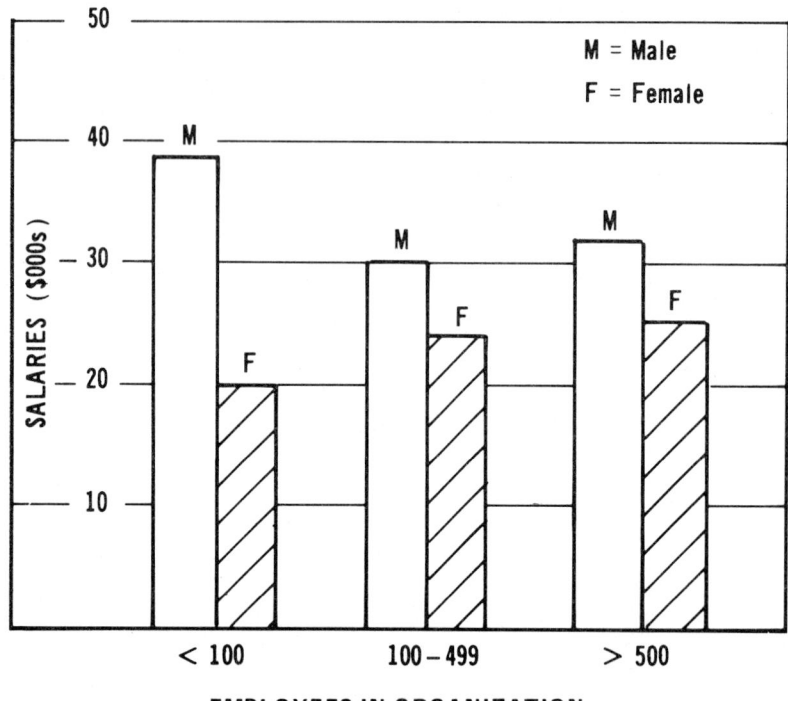

FIGURE 20 Annual earnings by size of establishment. SOURCE: "Software worker's survey," *Computerworld*, Nov. 14, 1983. © 1983 by CW Communications/Inc., Framingham, Mass. 01701. Reprinted with permission.

ence Foundation in 1981.[40] This suggests a disproportionate effort on their part to obtain grant support for their research. However, women received only 1 percent of the awards made that year. In the following year, women submitted only 1.8 percent of the proposals received by the same funding unit, but they received 1.5 percent of the awards. Entering faculty women had more success with Research Initiation Grants in 1983, submitting 3.7 percent of the proposals that year and receiving 4.7 percent of the awards.

The faculty member in a research-oriented institution who fails to attract grant support is seriously handicapped in pursuing a successful academic career. He or she is at a considerable disadvantage in attracting graduate students and hence in carrying out a viable research program. If women faculty have received less support, especially grant support, in the past than have their male colleagues, it should not be surprising to find them less productive in terms of the standard academic measures. There can even be a certain prophetic aspect to this. In the words of the sociologist J. R. Cole with regard to scientists in general: "If the initial assessment of who is apt to be a star is based on functionally irrelevant criteria such as gender, then the process of accumulating advantage can begin to enhance the career possibilities of men, while diminishing the chances of women."[42] Further, when women are defined as being less mobile geographically, which is often the case since women are more often than male colleagues one-half of a two-career couple, then this may reinforce the accumulation of career disadvantage.

The feelings of exclusion from informal collegial networks almost uniformly reported by faculty women in engineering and scientific fields suggest that women have not been accepted as colleagues with equal participatory rights in the informal activities associated with career success. This, too, can result in the continuing perception that academic opportunities are less than equal for women engineering faculty. If more women and members of minority groups are to be attracted into graduate school and ultimately into academic careers, the current environment in the universities needs to be examined and changed where it is found that the system discriminates unfairly in favor of white males.

Finding and Recommendation

The representation of women in engineering seems to be increasing at all academic levels. However, the same cannot be said in the case of minority groups. The presence of minorities in graduate schools is perceived to be principally a "pipeline" problem. Major efforts are needed with respect to minority groups at the junior high and high school levels, and upgraded retention programs are needed at the collegiate level. Efforts must be made to eliminate discrimination, real or perceived.

5
The Master's Degree

Master's degree graduates are well grounded in engineering fundamentals and design practices beyond the bachelor's level as the result of engaging in additional course work and practice. Often a thesis is completed, but by and large master's graduates are not likely to be as thoroughly capable of conducting independent research as are doctoral graduates. Ph.D.s are expected to have conducted original research in a specialized field of engineering, usually in addition to attaining the master's degree, and as a result are more experienced than are master's graduates. Furthermore, through their thesis experiences they are trained to discern, delineate, and solve problems.

The Council of Graduate Schools in the United States has described the master's degree in this way:[45]

Broadly speaking, the Master's degree indicates that the holder has mastered a program in a particular field sufficiently to pursue creative projects in that speciality. . . . The degree should be awarded for completion of a coherent program designed to assure the mastery of specified knowledge and skills, rather than for the random accumulation of a certain number of course credits after attaining the baccalaureate.

The Master's degree is customarily awarded to an aspirant who achieves a level of academic accomplishment substantially beyond that required for the baccalaureate degree. The Master's program should consist of a coherent pattern of courses frequently capped by comprehensive examinations and a thesis or its equivalent in a creative project. Ideally, all Master's programs should include an opportunity for the student to learn to present information in written and oral form to a variety of audiences.

A thesis has been a requirement for the Master's degree since its inception and has traditionally been a modest contribution to knowledge, certainly original to the student, and it may be original to the field. Although the thesis is not now a requirement in many Master's programs, a component demonstrating creativity should be required in quality programs.

With regard to the thesis, the *Manual of Graduate Study in Engineering* says:[46]

It is only at the doctorate level that there is justification for the requirement that a thesis shall comprise an original contribution to knowledge as evidence of expertise acquired. In respect to the thesis for a Master's degree, the time conventionally assigned is limited and the student is usually inexperienced in research. It should be clear, therefore, that the objectives of the Master's thesis are not necessarily the same as those of the doctorate. A commonly accepted principle in the curriculum leading to a Master's degree in engineering, is that the Master's thesis is to be primarily considered as a contribution to the training of the candidate rather than a contribution to knowledge.

According to the 1983 publication *Engineering College Research and Graduate Study*,[47] 196 U.S. institutions reported that they offer master's degrees, 150 of them with nonthesis options. In most cases where explicit information was given, it is clear that the nonthesis options include a requirement for a report or a project. In a few cases it was explicitly stated that no thesis, and presumably no project, was required. Stanford University, the nation's largest producer of engineering master's degrees (753 master's degrees in 1982–1983[48]) and one of the nation's most respected graduate schools, specifically states that a thesis is not required for a master's degree. On the other hand, the Massachusetts Institute of Technology, the nation's second largest producer (602 master's degrees in 1982–1983), does require a thesis.

Fifty-five of the 196 U.S. institutions reported that they offer the Master of Engineering degree. Most of these stated that a report, a project, or a "design problem" is required for this degree. Three institutions reported that they require industrial experience or an internship for the Master of Engineering degree. In most cases 30 semester hours (45 quarter hours) of work are required for the degree, making it nominally possible to be completed in one year.

Those who favor the elimination of the thesis as a requirement for the master's degree tend to feel that the numbers of master's theses each year are too great to permit adequate supervision, particularly if there is simultaneously a heavy load of doctoral theses. Hence, they worry whether the master's thesis requirement properly fulfills its role to provide a challenging, creative experience for the student. On the other hand, there are many who feel strongly that a master's thesis is an important "capstone" requirement and that it provides a major opportunity for a graduate student to undertake a project activity under his or her own initiative.

A significant, well-supervised master's thesis project clearly has value for the student, but the value declines if the thesis is allowed, under the press of other business, to deteriorate into a routine exercise. The thesis (or creative design project) should be structured in such a way as to be a meaningful, creative experience for the student. While it need not be an original contribution to the knowledge of the field, it should be sufficiently difficult to challenge the student's best capabilities and should be original with the student. It should not represent a project that is within the technical scope of a baccalaureate engineering education, but should call upon knowledge and skill that lie substantially beyond the usual baccalaureate level.

One of the purposes of a graduate program is to deepen a student's understanding of fundamental material. This, in turn, implies that there will be a greater emphasis on mathematics and science at the graduate level than at the undergraduate level, as well as an increasing degree of specialization. At the master's level, students will be moving in the direction of the "cutting edge" of technology, where less material has been formally reduced to textbook form and much information is in the process of being discovered and organized. Since the latter process is the one called research, it is natural that graduate education is inextricably interwoven with research. At the master's level the involvement with research will typically be less than at the doctor's level. In the case of doctoral programs, the students are expected to work directly at the frontier of knowledge, and, in fact, to make original contributions of their own to the body of knowledge. The intimate involvement with research is what characteristically separates undergraduate and graduate education. Undergraduate students may be involved with research on occasion, master's students usually will be, and Ph.D. students always are.

The utilization of master's degree holders in industry differs from company to company. In many companies they are employed to do the same kind of work that a bachelor's degree engineer does. However, those with a master's degree can make a contribution sooner and with less additional training and experience. In other companies the view is that the master's degree is indeed a more specialized degree than the bachelor's, and the recipients are employed on more technically demanding tasks, similar to but perhaps less demanding than those that challenge a doctoral recipient. Statistics from the Engineering Manpower Commission indicate that—including those earned by fully employed engineers who are part-time students—about one master's degree has been awarded for every three bachelor's degrees in recent years. This suggests that a fairly broad spectrum of utilization exists for the master's degree holder. But since this ratio has not been growing (in fact it has shrunk since the mid-1970s), it is clear that the master's degree is not becoming the "standard" entry-level engineering degree supplanting the bachelor's degree, no matter how attractive a goal that may seem to some.

Today's typical baccalaureate program provides reasonable coverage in mathematics, basic sciences, engineering science, and engineering design. Humanities and social sciences are required but seldom are sufficient to prepare the student for an effective role as a "totally educated person."

After provision has been made in the undergraduate engineering curriculum for the components mentioned above, the remaining portion (about one year) is intended to provide for additional breadth, or sometimes depth, in the general disciplinary area of study. In the discipline of civil engineering, for instance, this portion of the program is utilized to provide exposure, at a minimum level, to the several subdisciplinary areas—structures, water supply, waste disposal, transportation, geotechnical—as well as to develop some rigor in problem solving, design, synthesis, and planning. Unless the student is exceptional in that his or her goals are very clear, the possibility of incorporating any depth in a specific area is limited.

Most consulting engineering firms, regardless of the discipline, find that their practice requires the breadth described above but also requires additional depth in one or more of the subdisciplinary areas. While not requiring that all members of a team be expert in all of the related fields, a general understanding of the related elements is required of the participants in order for the problem-solving process to proceed effectively.

There is a dilemma in the practice of engineering by disciplines: on one hand, the problems presented are increasingly difficult, involving a much higher degree of sophistication—larger buildings, less desirable sites, energy conservation, dwindling material resources—thus requiring a more rigorous background and understanding of engineering science. On the other hand, these same problems are much more complex—hazardous wastes, environmental concerns, and economic and social aspects—so that their solutions require an interdisciplinary or multidisciplinary approach.

In the experience of many consulting firms, their success requires that additions to their professional staff be at the equivalent of a master's-level program. Generally, such a program would ideally be oriented toward substantial depth in one of the subdisciplines with emphasis on design practice or problem solving rather than on research or theory.

Industrial practice, while being more amenable to the development of specialized divisions in an engineering organization, also has many of the elements described above. This not only suggests a more general approach to the master's graduate program but recognizes the need for more thorough preparation for rapidly changing technology, with its implications for response to midcareer changes and adaptation. It may also come to imply the development of an advanced-level program that would include an internship in engineering practice.

Of particular concern is a recognition of the inadequacy of a single, concentrated educational experience as preparation for a lifelong career. Not only are today's graduates less likely to "stay the course" in one organization throughout their working lives, but today's organizations are equally less likely to remain in their current areas of involvement.

The graduate must be prepared to respond to change—the different requirements of a new employer or the changing requirements of a continuing employer. In the latter case, the change may require a substantial period of retraining at periodic intervals. The basic education must prepare the person to accomplish that task successfully.

The master's degree in some branches of the engineering profession has assumed the role of a "capstone" degree—the highest educational level to be sought, with no intention of proceeding to a higher degree. This produces a situation quite different from that prevailing in most fields of science, where full professional recognition by other scientists is usually accorded only to those with doctor's degrees.

In some fields of civil engineering and in most fields of electronics and computers, the master's degree has become, for the most part, the standard level of academic preparation for those engaged in design. As a result, a strong demand for part-time master's programs has come into being for engineers who are full-time employees in industry.

Several different modes of part-time master's programs have developed. The most common is the evening program; at some universities the graduate programs are almost entirely conducted in evening hours. A variant of the evening program is one that offers graduate classes in the early morning hours, from 7:00 a.m. to 10:00 a.m. (sometimes called early-bird programs). The advantage is that classes can be completed early in the day, since most business meetings and crises that might interfere with class attendance are not likely to occur before the midmorning hours.

TV has been employed by many universities to reach fully employed engineers, as was mentioned earlier. One popular mode uses "live" TV, and beams graduate courses to nearby industrial employers via closed-circuit microwave. A variant of live TV is the use of videotapes of courses, which are mailed, or sometimes carried by special courier, to the industrial locations.

Time saving is the principal advantage of TV systems, since the industrially employed students only have to go to where the TV sets are instead of traveling to campus. The principal disadvantage is loss of student–faculty contact. Offsetting this disadvantage, however, is the fact that fully employed students often display a high degree of enthusiasm for their part-time studies. The reason for this drive is presumed to be that such students, besides being more mature than usual graduate students, are likely to see a good "payoff" in taking courses closely coupled to their activities. Fully employed part-time students are generally so appreciative of the availability of TV courses that they usually say only positive things about such systems, even when the systems sometimes malfunction; on the other hand, full-time on-campus students often show extreme resentment for courses by TV, whether live or on tape. Unlike part-time off-campus students, they are likely to perceive every deficiency—image quality, system malfunction, inaccessibility of instructors—as a serious threat to the integrity of their educational programs.

Part-time degree-oriented programs can be included with other kinds of course work, either with or without credit, under the general heading of *continuing education*. Since continuing education is the subject of a study being done parallel to this one, it is not treated extensively here. However, the subject of technical obsolescence is discussed briefly in the context of continuing education.

"Obsolescence," for engineers, can occur in many different ways. Rarely, however, does it occur in an individual who remains active in a given field, with respect to the knowledge content of that field. Such an individual generally becomes more rather than less capable in his or her field. But obsolescence can occur quickly as a result of external factors. For example, an engineer's field can become of little or no interest to the engineer's employer if the "action" moves to a different field. This happened when vacuum tubes declined in importance and were largely supplanted by transistors. The charged particles in evacuated media continued to behave in the way they used to—such laws remained as valid as they ever were; but vacuum tubes were not used much any more. It didn't help that vacuum tube engineers were more competent than ever in their chosen field—it was still necessary for most of these engineers to move into a new field, or perhaps to apply the knowledge of the behavior of charged particles to another field where such knowledge was applicable, such as plasma physics.

"Obsolescence" can also occur if fundamental knowledge that was once well understood by the engineer has been largely forgotten because of nonuse. Many engineering schools require their mechanical and civil engineers to take courses in electric circuits, for example. But if this material is not used it will be forgotten. It may be possible for the engineer to recapture the forgotten material through independent study, but individuals often seek out formal courses instead.

Companies offer extensive amounts of in-house education to their engineers, often because of the companies' movement into new technologies. The widespread advent of computers and of computer-aided manufacturing has required a great deal of continuing education, much of it in-house. So has the arrival of the microprocessor, which has completely changed the way many products operate. Such courses obviously help ward off the spectre of engineer obsolescence; yet, employers worry about obsolescence and how to combat it. For example, an extensive study of policies in 17 research and development laboratories showed that there was a general fear that organizational productivity would decline as the average age of their technical staffs increased.[49]

Although it is widely thought to be so, there seems to be little hard evidence that productivity truly does decline with age. More likely, "obsolescence" is the result of field shifting or of forgotten unused material. In fact, in the study of R&D organizations cited above, there is some evidence that scientific productivity may actually increase after age 50, although the productivity may consist more of things like pulling together the ideas of one's life work than of coming up with major new ideas. However, in another study, the investigators *did* find evidence of declining performance with age, based on evaluation of engineers by their managers.[50] An especially significant finding in this study was that the routine taking of courses for the purpose of continuing education seemed to have no effect on performance. The investigators found another result that they thought was of special importance: engineers with advanced degrees were considered productive up to 10 years longer than those with bachelor's degrees. Hence, the investigators made the recommendation that midcareer graduate work inten-

THE MASTER'S DEGREE

sive enough to result in a degree might effectively prolong an engineer's productive life. However, one could legitimately question whether it is the graduate program per se that has such a beneficial effect; it could just as well be argued that the personal drive of the individual is the factor that holds off obsolescence and also causes him or her to enter a protracted and intensive program leading to a graduate degree.

Earning a master's degree can be quite a flexible matter, since many schools make it possible to complete a master's degree in one year. Because of the wide availability of part-time programs, an engineer frequently can make the decision to go into a master's program without simultaneously facing the decision to resign from full-time employment. Maintaining access to such programs is important to many engineers. Universities should provide evening courses or should utilize technologies such as TV courses, live or videotaped, to reach fully employed engineers in industry.

It is useful to know something of the attitudes of engineers toward graduate study. A survey of 3,246 engineers was conducted in 1964–1966 as a part of the Goals Study.[51] The response of the engineers to a set of questions regarding graduate study is shown in Table 22. Roughly half of the respondents felt that graduate work was needed. In a 1981 survey of 3,401 engineers, a similar set of questions was asked, with the responses also shown in Table 22.[52] In the 1981 study, responses were tabulated by sex of the respondent; Table 22 reveals virtually no difference in the attitudes toward graduate study between male and female engineers.

In 1966, 57 percent of the respondents thought graduate work in management was important; in 1981, this percentage was about 50. In 1966, 52 percent of the respondents thought graduate work in mathematics and science was important; in 1981, this percentage was about 30, but 47 percent of the 1981 respondents thought graduate work in *engineering* was important.

Table 22 shows that the respondents were divided about the particular form that continuing education should take, but it is worth pointing out that continuing education takes many forms, not only that of lecture courses. In the 1981 study just cited, the respondents reported that they had engaged in the following activities during the preceding year:[52]

Activity	Respondents
Discussed new engineering developments	68%
Read about new engineering developments	79
Subscribed to engineering periodicals	79

(Continued on p. 81)

TABLE 22 Opinions of Professional Engineers Regarding Graduate Study

		Percent Who "Strongly Agree" or "Agree"		
			1981	
	1966	All Respondents	Men Respondents	Women Respondents

Q. Below are some statements about the need for graduate work or continuing education in your present field of employment. For each statement indicate the extent to which you agree or disagree.

	1966	All Respondents	Men Respondents	Women Respondents
1. A bachelor's degree is sufficient preparation; graduate study is not needed.	53	59	59	59
2. Graduate degree engineers perform different functions than bachelor degree engineers.	42	—	—	—
3. Formal instruction in modern developments in technology is necessary for keeping "up-to-date."	63	—	—	—
4. On-the-job training or "in-house" courses are sufficient for keeping "up-to-date."	—	47	47	46
5. Noncredit courses are sufficient for keeping "up-to-date."	—	56	56	56
6. Graduate work is needed, but in management.	57	50	49	51
7. Graduate work is needed in math and science.	52	31	32	30
8. Graduate work is needed with emphasis on engineering.	—	47	48	46
9. Graduate work is needed, but with emphasis on broadening my knowledge in other fields of engineering.	51	—	—	—

SOURCES: W. K. LeBold et al., "The Engineer in Industry and Government," *Engineering Education*, March 1966, pp. 237–273; W. K. LeBold et al., *National Engineering Career Development Study* (West Lafayette, Ind.: Purdue, June 1983).

THE MASTER'S DEGREE

(Continued from p. 79)

Activity	Respondents
Read new books on engineering or science	40
Purchased new books on engineering/science	40
Attended local technical meetings	46
Took nongraduate credit engineering course	16
Completed graduate courses in engineering	15
Attended national technical meeting	28
Presented one or more technical papers	11
Attended short course on management	28

It is possible that the activities "discussed new engineering developments" or "subscribed to engineering periodicals" may not represent much of a commitment to continuing education, but "read new books on engineering or science" (40 percent) and "attended national technical meeting" (28 percent) represents a heavier commitment, as does the taking of courses, whether in engineering or management. In particular, if "took nongraduate credit engineering course" and "completed graduate courses in engineering" are nonduplicative, then 31 percent of the respondents engaged in formal course work in engineering.

There were significant differences in the makeup of the 1966 and 1981 pools of respondents. The 1981 pool had been deliberately selected to consist mostly of young engineers: 71 percent of them had less than 10 years of experience. In the 1966 pool a substantially smaller fraction was in the young group: only 40 percent had less than 12 years of experience. The educational levels of the two groups were as follows:

	1966	1981
No degree	1%	1%
Bachelor's degree	71	56
Master's degree	17	35
Doctor's degree	11	6
Other	—	3

Furthermore, the 1981 group consisted of approximately 41 percent women while the 1966 group was virtually all male. In spite of the differences in the makeup of the groups and the elapsed time between 1966 and 1981, the responses were strikingly similar: about half the engineers in both surveys thought graduate work was needed, and gave their votes in about equal proportions to graduate study in engineering or in management.

Findings and Recommendations

1. Well-supervised master's thesis projects have great value for students, but their value becomes questionable if they are allowed to degenerate into routine exercises. A master's thesis need not represent an original contribution to the field, but it should be original with the student. It should represent a meaningful, creative experience from the student's point of view.

2. A single, concentrated educational experience is not sufficient for a lifelong career. Graduates must be prepared to respond to the changing requirements of their employers, which may require retraining at periodic intervals. Universities should provide programs for such a purpose, either through TV or other means. Employers should provide in-house educational programs.

6
The Doctor's Degree

The training of the Ph.D. engineer has been the subject of debate for many decades and is of special current interest because of the shortage of Ph.D.s in engineering for faculty positions. While there is a great need for Ph.D. engineers in academia, there are not equally well defined needs for large numbers of Ph.D.s in all industries. The Ph.D. engineer is generally utilized effectively by industry in research or development; however, that same degree and type of training are much less needed in operational divisions. While having a Ph.D. is advantageous in business, it is by no means a requisite for success.

Industrial research is being driven by major advances taking place today in science, in analytical tools, and in computing capability as well as in market sophistication. Highly trained doctoral-level engineers are needed to tackle these complex research and development challenges. These experts command high starting salaries in industry relative to academia and are often given challenging assignments with access to the most modern facilities and equipment.

Regarding the nature and purpose of a doctoral program, the Council of Graduate Schools in the United States has stated:[53]

> The doctoral program is designed to prepare a student for a lifetime of intellectual inquiry that manifests itself in creative scholarship and research, often leading to careers in social, governmental, business, and industrial organizations as well as the more traditional careers in university and college teaching. The program emphasizes freedom of inquiry and expression and development

of the student's capacity to make significant contributions to knowledge. An essential element is the development of the ability to understand and evaluate critically the literature of the field and to apply appropriate principles and procedures to the recognition, evaluation, interpretation, and understanding of issues and problems at the frontiers of knowledge.

A dissertation is universally required in U.S. universities for the doctor's degree. The purpose of the dissertation is twofold: (1) to develop in the candidate the independent ability to carry out a scholarly investigation of a challenging topic at a high level of professional ability, and (2) to provide for an original contribution to knowledge in the field. Generally, the candidate is expected to defend the dissertation in a final examination; sometimes such examinations are open to the public. A nearly universal doctoral requirement is a comprehensive examination consisting of written and oral parts, generally imposed just before the candidate begins work on the dissertation. The purpose of the examination is to demonstrate an adequate knowledge of the field and an ability to use academic resources. If the candidate passes the examination, it is considered likely that he or she will successfully complete the dissertation. Many schools additionally impose yet another examination, given early in the student's program, to determine fitness for doctoral work.

The foreign language requirement tends to vary from school to school and frequently from department to department within a given school. Some kind of foreign language reading ability at one time was a nearly universal requirement. Now many departments have no foreign language requirement, apparently in the belief that such requirements generally produce little utility in reading articles published in foreign languages. The policy statement on the Ph.D. degree by the Council of Graduate Schools does not specify a foreign language requirement, and in fact scarcely mentions the topic.[53]

Graduate schools generally require two years *in residence* for doctoral programs, in order to provide an appropriate degree of student–faculty interaction and supervision of the thesis research. The result of the residence requirement is usually that a candidate must forgo full-time employment and become a full-time student. However, fully employed individuals have sometimes been able to complete doctoral programs without giving up their employment if they are close to a university campus and if their job assignments are flexible enough to permit extensive student–faculty interaction and faculty supervision of the thesis work.

Since Ph.D. programs generally require a minimum of two years in residence and master's programs require a minimum of one year, the minimum elapsed time between the B.S. and Ph.D. degrees for a student enrolled in full-time study would be three years. However, data collected by the National Research Council show that the average "registered" time between these two degrees for engineers is 5.8 years, and the average total elapsed time is 7.5 to 7.9 years.[14,15] These periods of time are greater than the minimum time, because few graduate students proceed directly from the baccalaureate to the doctor's degree with unbroken full-time study. There is often a period of employment before graduate school; also, doctoral students are frequently employed on a part-time basis by their universities as postgraduate researchers or as teaching assistants; furthermore, students often are required to take course work beyond the minimum in order to fill in gaps in their backgrounds; and, finally, there is the classic reason, which is that the thesis takes longer than expected.

According to *Engineering College Research and Graduate Study*, there were 149 U.S. universities offering doctoral engineering programs in 1983.[47] Of these, 137 reported that they awarded one or more doctor's degrees in 1982–1983.[48] Nearly half (46 percent) reported fewer than 10 doctorates each, and 26 percent reported fewer than 5. The 30 largest engineering doctorate producers are listed in Table 23. In 1982–1983, these 30 institutions produced 62 percent of the engineering doctorates in the country. Table 23 also shows that the ratio of doctorates per faculty ranged from 0.12 to 0.75 in 1982–1983, with an average of 0.32. It is likely that many of these schools could increase their Ph.D. output, although in order to do so they would probably have to reduce their undergraduate loads or else expand their resources, because most schools already are overloaded. It would seem reasonable that many of the schools which presently have small Ph.D. outputs could expand their production, and that they have an adequate quality base from which to do so. The conclusion, then, is that our existing system of engineering graduate schools is capable of expanding its production to the needed levels and that the startup of additional Ph.D. programs should not be encouraged. Expanding on the present base will require additional faculty and other resources, but is less expensive than starting new programs.

Doctoral programs must have a strong base of funded research. Table 24 provides some insight to the magnitude of the cost. The total 1982–1983 research expenditures for the 30 institutions listed in Table 23 are shown in Table 24, and are broken down into the following categories: (1) federal government, (2) state and local government, (3) business and industry, and (4) an "other" category, which includes private nonprofit organizations and institutional support. Not all universities reported

TABLE 23 Thirty Largest Engineering Doctorate Producers (based on average of 4 years, 1980–1983)

		1982–1983		
School	Degrees per Year (4-yr avg)	Degrees	Faculty	Degrees/ Faculty
MIT	166	171	373	0.46
Illinois-Urbana/Champaign	156	145	392	0.37
Calif.-Berkeley	144	170	231	0.74
Stanford	124	127	169	0.75
Purdue	94	88	280	0.31
Cornell	74	79	208	0.38
Calif.-UCLA	66	50	139	0.36
Michigan	66	63	257	0.25
Northwestern	62	64	127	0.50
Wisconsin	57	58	192	0.30
Ohio State	56	64	309	0.21
Southern Cal.	52	55	142	0.39
Texas	47	55	176	0.31
Cal. Tech.	45	51	80	0.64
Rensselaer	44	63	163	0.39
Calif.-Davis	42	44	110	0.40
Iowa State	38	41	309	0.13
Harvard	37	30	51	0.59
Virginia Tech.	37	50	265	0.19
Carnegie-Mellon	36	37	105	0.35
Georgia Tech.	36	51	271	0.19
Penn State	36	41	282	0.15
Princeton	36	38	75	0.51
Texas A&M	35	40	334	0.12
Columbia	34	35	108	0.32
Minnesota	34	37	183	0.20
Univ. of Pa.	33	29	99	0.29
Case Western Reserve	32	36	111	0.32
Polytechnic Inst. of N.Y.	32	26	132	0.20
Colorado State	31	33	113	0.32
		1,871	5,786	0.32

SOURCES: *Engineering and Technology Degrees* (New York: Engineering Manpower Commission, 1979, 1980, 1981, 1982, 1983). *Engineering College Research and Graduate Study, Engineering Education,* March 1984.

data in these categories, and in other cases the data did not match the totals given for the entire engineering school; in such cases, N/A ("not available") is shown.

The total 1982–1983 research expenditure for each school has been divided by the average number of doctor's degrees granted per year in the 1980–1983 period to give a general idea of the research base supporting the doctoral enterprise. These ratios are shown in Table 24, with public and private institutions listed in different columns. These "dollars per degree" numbers should *not* be considered as the cost per degree and should not even be taken too literally. There are too many differences in the modes of operation of the different schools to permit such a simple interpretation. Engineering schools have, for example, varying kinds of organized research units with various organizational relationships to the schools. Some may be wholly within the schools; others may be partially or wholly outside the engineering school organizations, and even though the research of the doctoral students may be heavily supported by these research units, the schools may report the dollar amounts in different ways. Furthermore, for some schools a significant portion of their research contracts may have little involvement on the part of doctoral students. Hence, the figures shown in Table 24 should be taken as the most general of guides concerning the dollar magnitude of research required to support a major increase in doctoral output.

The columns headed "$ per degree" in Table 24 show an average of $241,000 per doctoral degree for public institutions and $317,000 per degree for private universities. Thus, if we conservatively use a figure of $200,000 per doctoral degree, we could anticipate that an increase in funded research for engineering on the order of $200 million per year would be required to support an increase from the present level of 3,000 Ph.D.s per year to the projected level of 4,000 or so per year by 1988. The last column of Table 24 shows that only a small fraction of the total research funding—on the order of 15 percent—has historically been provided by industry. Thus, unless there is a very large increase in research funding by industry, most of the increase will have to be provided by government, and this means the federal government primarily, if past patterns prevail. Provision will need to be made also for major upgrading in research equipment, coming in part from the research contracts themselves.

A different approach to this same topic may be taken by examining research support data for the Engineering Directorate of the National Science Foundation. In Fiscal Year 1983, NSF supported 1,795 graduate students for a research dollar total of approximately $101 million. Thus, there was an average of approximately $56,300 of support per graduate student. Of this, only about $9,700 went directly to the salary for each graduate student, on the average, with the rest in salaries for faculty, postdoctoral students, undergraduate students, secretaries, technicians, and costs for equipment, travel, computer time, supplies, and general overhead. The 1,795 graduate students consist of a mix of

TABLE 24 Research Expenditures in Engineering Schools, 1982–1983 (30 U.S. Universities)

	Doctoral Degrees (4-yr avg)	Total	Federal Government	State and Local Government	Business and Industry	Private Nonprofit Organizations, Institutional and Other	$ per Degree Public Universities	$ per Degree Private Universities	Business and Industry $ per Degree
MIT	166	$53,592,000	N/A	N/A	N/A	N/A		$323,000	—
Illinois-Urbana/Champaign	156	36,058,000	$28,094,000	$ 3,120,000	$ 2,112,000	$2,733,000	$ 231,000		$ 13,500
Calif.-Berkeley	144	25,905,000	16,354,000	4,600,000	4,260,000	691,000	180,000		30,000
Stanford	124	37,000,000	34,133,000	34,000	831,000	2,002,000		298,000	7,000
Purdue	94	23,334,000	12,798,000	748,000	5,325,000	4,463,000	248,000		57,000
Cornell	74	27,101,000	23,412,000	—	3,306,000	383,000		366,000	45,000
Calif.-UCLA	66	12,584,000	10,217,000	5,000	1,736,000	626,000	191,000		26,000
Michigan	66	16,852,000	14,048,000	3,000	1,968,000	563,000	255,000		30,000
Northwestern	62	9,500,000	N/A	N/A	N/A	N/A		153,000	—
Wisconsin	57	17,291,000	10,048,000	2,418,000	3,167,000	1,658,000	303,000		56,000
Ohio State	56	18,898,000	9,534,000	1,662,000	4,565,000	3,137,000	337,000		82,000
Southern Cal.	52	23,295,000	23,057,000	—	238,000	—		448,000	5,000
Texas	47	13,395,000	7,461,000	2,605,000	1,434,000	1,895,000	285,000		31,000
Cal. Tech.	45	13,741,000	9,445,000	N/A	N/A	4,296,000		305,000	—
Rensselaer	44	19,332,000	8,797,000	148,000	7,508,000	2,879,000		439,000	171,000
Calif.-Davis	42	5,447,000	2,895,000	1,877,000	675,000	—	130,000		16,000

Institution									
Iowa State	38	6,366,000	1,400,000	471,000	822,000	3,673,000	168,000		22,000
Harvard	37	9,779,000	8,679,000	—	594,000	506,000		264,000	16,000
Virginia Tech.	37	10,752,000	5,159,000	4,285,000	1,112,000	196,000	291,000		30,000
Carnegie–Mellon	36	16,845,000	10,117,000	—	5,527,000	1,201,000		468,000	154,000
Georgia Tech.	36	71,440,000	38,578,000	20,003,000	11,430,000	1,429,000	1,984,000		318,000
Penn State	36	9,202,000	5,580,000	790,000	2,440,000	392,000	256,000		68,000
Princeton	36	8,535,000	N/A	N/A	N/A	N/A		237,000	—
Texas A&M	35	34,178,000	9,528,000	21,669,000	2,546,000	435,000	977,000		73,000
Columbia	34	10,420,000	8,040,000	—	2,213,000	169,000		306,000	65,000
Minnesota	34	8,286,000	5,731,000	203,000	1,992,000	360,000	244,000		59,000
Univ. of Pa.	33	11,497,000	10,037,000	77,000	677,000	706,000		348,000	21,000
Case Western	32	11,740,000	9,534,000	N/A	N/A	N/A		367,000	—
Polytechnic Inst. of N.Y.	32	3,642,000	N/A	N/A	N/A	N/A		114,000	—
Colorado State	31	13,898,000	9,086,000	881,000	3,673,000	258,000	448,000		118,000
Average							$ 241,000[a]	$317,000	$ 41,000

NOTE: N/A = not available.

[a] Georgia Tech and Texas A&M not calculated into this average because their values per degree ($1,984,000 and $977,000, respectively) were so markedly different from the others.

SOURCES: *Engineering and Technology Degrees* (New York: Engineering Manpower Commission, 1979, 1980, 1981, 1982, 1983). *Engineering College Research and Graduate Study, Engineering Education*, March 1984.

master's and doctor's students, but a comparison with the figure derived earlier can be made as follows. Let us suppose all of the 1,795 students are doctoral students, and that the students are evenly distributed throughout the four or five years that are required to complete a doctoral program. We can then estimate the number of doctor's degrees *per year* at one-fourth or one-fifth of the total in the program. If we take the more conservative figure of one-fourth, then a program with 1,795 doctoral students would produce about 450 Ph.D.s per year. Dividing 450 degrees per year into the annual expenditure of $101 million gives a figure of $224,000 of support per Ph.D. degree per year, which correlates well with our earlier figure. Using the assumption of five years to the degree would give a figure of $280,500 per Ph.D. degree per year.

The available base of facilities and equipment has fallen well behind the needs in the face of overall enrollment growth. A specific statement concerning the total need is not possible in the absence of a detailed nationwide inventory of needs and resources. However, it should be noted that space needs per student are greater for graduate than for undergraduate students. The principal need in the case of undergraduates is for classrooms and class laboratories, with related support needs such as computer facilities and shops. For graduate students, in addition to classrooms and support facilities there is a substantial need for research laboratories. For one research university, the University of California, the space standards provide for 200 square feet of laboratory space per graduate student, and 300 square feet of lab space per faculty member, plus another 220 square feet of office and clerical support space per faculty member. If we made the assumption that all available space nationally is already utilized and that there is no surplus available, then the current projection of growth in Ph.D. production by 1,000 per year (implying 4,000 additional students registered), plus the space needed to meet the full shortfall of 6,700 faculty, would require approximately 4.5 million square feet of new space as an upper limit. If space costs $100 to $250 per square foot, depending upon the sophistication of the laboratories, then the estimated cost ranges from $450 million to $1 billion (1983 dollars) nationwide. For the states with the largest engineering enrollments (California, Texas, and New York) on a proportional basis, this could require an investment on the order of $60 million to $80 million per state for expansion of facilities and basic equipment.

There appears to be no definitive data base regarding the true magnitude of the need for research equipment in engineering schools. In Fiscal Year 1984, the National Science Foundation budgeted approximately $18 million for equipment, out of its total engineering budget of

$123 million.[62] NSF accounted for 38 percent of all federal obligations to universities and colleges for basic research in engineering in Fiscal Year 1983,[63] so the total federal support for engineering research equipment probably lies in the range of $50 million to $75 million per year. NSF has estimated that it has been able to service only a small fraction of the total need, as evidenced by equipment requests in proposals. A part of the need can be met by the $200-million increase in funded research recommended earlier. This could amount to $35 million to $40 million per year if past patterns prevail. An additional portion will be met by the $450-million to $1-billion construction estimate set forth in the preceding paragraph, since construction budgets typically include some provision for built-in equipment. The unmet equipment need could easily be on the order of $100 million, but this figure cannot be supported by any straightforward analysis. Since much of the need would be met by the construction budget discussed earlier and by an augmentation of engineering research of $200 million per year, no separate dollar recommendation for this element is included here.

The major increases in federal funding needed to support increased doctoral programs would no doubt come from a variety of government agencies. The traditions of the National Science Foundation have been productive for the sponsorship of academic research in engineering, although other government agencies have also found effective ways to sponsor academic research within their operating guidelines. Increased emphasis on engineering research within the National Science Foundation, a process that is already occurring, would be a strong stimulus toward the objectives outlined in this report.

An important issue for engineering schools is the fact that research leadership in some fields has shifted substantially from academia to industry, as in the cases of VLSI and automated manufacturing based upon CAD/CAM. Such shifts are especially likely to occur in fields where the costs of laboratories are so great that few universities can afford them. The seriousness of this situation for American technical education is that graduate work should be couched in a research environment at the cutting edge of technology, and if the cutting edge is in industry, the educational experience will be less valuable than it should be.

For a handful of universities, industrial funds have been brought together to establish major research facilities in specialized fields. Industry should be encouraged to provide such support to the maximum extent feasible, and it should seek to support a mix of fields rather than a narrow selection. Government support has also been brought to bear on the problem of up-to-date research facilities in universities. All of these measures are beneficial in helping to establish cutting edge research environments in universities, but they cannot cover all the needs. Additional measures that can be taken include cooperative industry–university research and consulting relationships for faculty. These have great potential benefits for academic institutions because they bring the academic environment closer to the moving frontier of industrial practice.

Findings and Recommendations

1. The existing system of engineering graduate schools should be capable of expanding its doctoral production to the increased level that is needed, and the startup of additional Ph.D. programs should not be encouraged.

2. An increase of doctoral output will entail a corresponding increase in funded research. It is estimated that an increase on the order of $200 million of new funded research per year will be required, principally from the federal government.

3. The available base of facilities and equipment has fallen well behind the needs for engineering education. Expansion of the Ph.D. output, plus meeting the needs of the full 6,700 "shortfall" in faculty, would require space on the order of 4.5 million square feet of new space as an upper limit. The upper-limit cost of such space, depending upon the sophistication of the laboratories, would range from $450 million to $1 billion nationwide (1983 dollars).

4. The traditions of the National Science Foundation represent an excellent model for funding university research. Increased emphasis on engineering research within the National Science Foundation is strongly encouraged.

7
Nontraditional Graduate Programs

The concern has occasionally been expressed that engineering graduate programs are too theoretical, too much concerned with research and not enough with design. In particular, the view is held by some that the typical new Ph.D. has not had an educational experience that is readily adapted to industry's needs. The criticisms can be reduced to two: (1) The Ph.D. graduate is too analytical, too abstract, and too inclined to believe that a solution is not a worthy one unless it is mathematically elegant. (2) The prolonged period of hard work required to develop expert competence in a specialized area results in an unwillingness to abandon that hard-won position to work in a different area, an area that may happen to be of interest to the engineer's employer.

The response of some institutions has been to design educational programs that meet criticism (1) by requiring a "design" emphasis, or (2) by eliminating the research dissertation. Programs resulting in the degree "Doctor of Engineering" (D.Engr.) are typical of the former, and programs awarding the degree "Engineer" are typical of the latter.

A matter of some concern is the organizational framework in which advanced "professional" degrees are to be embedded. In the eyes of some, graduate engineering programs cannot develop a proper professional flavor unless they are under the separate jurisdiction of the engineering school and, therefore, are free of control of a campuswide graduate school. However, in a survey of universities that offer the Doctor of Engineering or the Engineer degree or both, few respondents made a point of this;[54] most universities with these degrees apparently administer them through the graduate school.

In a 1972 survey, Lawrence N. Canjar found that most engineering schools did not consider the matter of separate jurisdiction for engineering graduate degrees to be important. He stated, "It was hoped that . . . jurisdiction over professional curricula and professional degrees might be used to separate out the professional from the purely academic programs, but the questionnaire indicated that 90% of the institutions responding felt secure in the operations of their graduate programs."[55] L. E. Grinter, in a 1975 article, offered five criteria that could be used to define a professional school of engineering, but the article did not even mention separate jurisdiction over the graduate degrees.[56] The impression created is that the particular organizational structure is not a major factor governing the nature of the degree.

The Doctor of Engineering

According to the March 1983 issue of *Engineering Education*, entitled *Engineering College Research and Graduate Study*, 16 U.S. universities offer the degree Doctor of Engineering.[47] Nearly all the schools offering this degree state that the D.Engr. dissertation differs from that of the Ph.D. in that it is more applications- or design-oriented than directed toward the development of basic knowledge. In the words of one university catalog: "The principal criteria of achievement in the dissertation are originality and creativity in the application of engineering tools to solve a significant and specifically defined problem."

In some cases, schools offering the Doctor of Engineering degree encourage some management or other nonengineering work in the total program. This is especially the case with Texas A&M, where D.Engr. students are required to take seven courses in management, ethics, and legal relationships in addition to the usual graduate-level engineering courses. The University of Kansas has a similar arrangement: D.Engr. students are required to select from courses in the areas of finance, marketing, organizational theory, and technology and society. Other schools may not actually require students to take such courses, but may encourage them to develop a minor from courses in law, business, economics, psychology, and political science. In programs that include a business minor, whether required or optional, the avowed intent is to prepare industrial leaders. According to one university, "Through this program we are attempting to turn out leaders, men and women who will start their professional careers as engineers, but who will be prepared to move into management and eventually into the upper ranks of industrial organizations."

The question of a required internship inevitably arises in connection with Doctor of Engineering programs. The rationale for an internship usually is that it gives the student an opportunity to apply his or her training to a practical problem, and provides exposure to a realistic industrial setting.

It would appear that the practice of requiring an internship has been borrowed from medical schools. But it seems to have been overlooked that an internship typically is not a requirement for the M.D. degree, but takes place after the degree has been awarded. In medical education a period of service as a "resident" is also generally required before full certification is granted. Together, the internship and residency constitute a period of beginning medical practice, supervised by clinical faculty who are themselves professional practitioners.

Most of those who favor internships for engineers generally believe that the internship should not come at the end of the formal academic program. It is argued that a year in an internship sandwiched between two periods of academic study has greater ultimate value to a young engineer than does the combination of a straight academic program followed by the first year on the job. The experience on the job is believed to give a special meaning and force to the following period of formal study. No doubt this is true, but it has also been argued that a young engineer will learn more during the first year of a job that is known to be the beginning of a lifetime career than would typically be the case in a year of internship, which is in a temporary job and may merely be looked upon as fulfillment of an academic requirement. Both sides of this issue have adherents, and it does not appear to be possible to settle the matter definitively.

Even though several schools have gone to considerable lengths to develop Doctor of Engineering programs, the number of students who select them has been small. Since the time and effort required of the student are essentially the same as for the Ph.D., most students apparently prefer the better-known degree.

Some have questioned whether the need exists for a "professional" doctor's degree separate from the Ph.D. After all, they say, over the years thousands of Ph.D. engineers have succeeded very well in professional roles for which the "research" doctorate is supposedly ill-suited. A comment from one university sums up a view that is often heard: "The Master of Science and Doctor of Philosophy degrees are administered through the Graduate School, but we have a considerable degree of flexibility as to the professional practice versus the research emphasis in the content of these degree programs. As a result, we believe our Master's and Doctor's degrees may be tailored to suit the professional practice interests of both the graduate students and engineering employers."

The Engineer Degree

The former practice of awarding the Engineer degree on the basis of a certain number of years of experience, plus a thesis, has nearly disappeared, and the usage of the degree as an intermediate one between the master's and doctor's degrees has taken its place. MIT, Stanford, and the University of Southern California award significant numbers of Engineer degrees each year, and a total of 21 institutions indicated, according to the 1983 issue of *Engineering College Research and Graduate Study*, that they offer the Engineer degree or equivalent.[47] The general view of this degree is as stated by one university: "An intermediate degree is necessary for those who desire to proceed beyond the M.S. but are not interested in the 'researchy' nature of the Ph.D. Stanford remains with the Engineer degree as the intermediate one for historical reasons and because the package seems to be a coherent and reasonable one."

The University of California at Los Angeles (UCLA) has a rather unusual arrangement regarding the Engineer degree. Its literature states: "The Ph.D. and Engineer Degree programs will be administered interchangeably in the sense that a student in the Ph.D. program can exit with an Engineer degree or even pick up the Engineer degree on the way to the Ph.D. degree, and similarly, a student in the Engineer degree program can continue for the Ph.D. after receiving the Engineer degree." UCLA established its Engineer degree in 1976 in response to urging from industry that a special degree program was needed to provide a period of study more advanced than the master's but less extensive than the doctorate. By 1982, 135 individuals had received the degree. However, about 90 percent of these had continued on to the Ph.D., so the program apparently had not served its intended function as a terminal degree sought for its own sake.

A delicate question surrounds the Engineer degree: Is it sometimes used as a consolation prize for those who flunk their doctoral exams? Most schools answer this question in the negative, although a few admit that the degree sometimes is awarded to those not considered fully qualified for a Ph.D. degree. However, this is certainly also the case with master's degrees, but one practically never hears of the master's degree, at least in engineering, referred to as a consolation prize. In view of the fact that Engineer degree programs seem to be increasing, perhaps the question will come up less often. This will be especially true if industry in a major way begins to seek the degree for its own sake, which does not yet seem to be the case.

Finding and Recommendation

Experimentation with new degree programs is a necessary activity in engineering education, and flexibility in existing programs is to be encouraged. However, quality of all programs should be continuously monitored to ensure that the technical competence of persons earning advanced engineering degrees is maintained at a high level.

8
The Engineering Faculty

For a number of reasons it has been difficult to attract and retain faculty members in U.S. engineering schools during recent years. A survey published in 1983 by the American Society for Engineering Education showed 1,400 unfilled faculty vacancies.[23]

The forces that caused an insufficient number of faculty to be available are principally these:

1. Too few Ph.D.s have been produced in recent years.
2. Faculty salaries have been noncompetitive with industrial salaries by too large a margin.
3. The availability of state-of-the-art research equipment in most universities has declined relative to industry, causing many potential faculty members to choose industrial careers.
4. Other factors have made faculty careers less attractive than those in industry, such as excessive teaching loads, lack of support for faculty travel and supplies, inadequate support staff and teaching assistants, and lack of space for development of both instructional and research activities.

The vitality of engineering education depends on maintaining a high-quality faculty consisting of an appropriate mix of persons with varying backgrounds and strengths. Some faculty should have their activities focused on the cutting edge of theory; others should concentrate their talents on the experimental verification of theory, and on the kind of theory–experiment interaction that characterizes the very best research. Still other faculty should engage their efforts in the region where theory becomes transformed into new professional practice, and yet others should concentrate on the teaching of the best current practice. Finally, a small portion may focus on other matters related to the practice of engineering, such as ethics, social impact, management, and law.

Not all schools will have faculty representing all of the facets listed above, although large schools may well have representation from all of them. Schools that emphasize undergraduate education will probably have their greatest strength concentrated in the teaching of current practice. Research universities have this function too, but will focus their strongest efforts on the development and verification of theory, coupled with the orderly transformation of this theory into practice, principally through the writing of monographs and textbooks.

In all of these cases, the faculty must have adequate opportunities for interaction with students. The teaching of engineering requires modeling the problem-solving process, practicing problem definition, nurturing the skill of making assumptions, providing feedback to the students as they progress through their solutions, and having the students evaluate their solutions once they are obtained. Students need to learn where their reasoning has gone wrong and why their assumptions are not reasonable, a need that applies to both graduate and undergraduate students.[22]

The Need for the Doctor's Degree

It has often been pointed out that not all faculty in an engineering school need to have doctor's degrees. Certainly, schools that emphasize programs in engineering technology have not demanded doctorates for all their faculty. The same can be said for schools specializing in undergraduate programs. Even research universities make a practice of employing a certain number of adjunct and part-time faculty to enrich the professional practice side of their programs, for which having a doctorate may not necessarily be a factor. These individuals will instead be chosen principally for their modern expertise and for their teaching ability. It is also a common practice in universities with graduate programs to employ their better graduate students to teach certain basic courses. These individuals will not have doctorates, and experience shows that they are sometimes among the very best instructors in the institution.

For universities with graduate and research programs, the percentage of faculty who can function in roles such as those described above, without doctor's degrees, will be small. In the great majority of cases the faculty will be guiding graduate students, will be teaching courses at or close to the cutting edge of new knowledge, and will be personally engaged in research or other kinds of creative activity. For them, the kind of background represented by a doctor's degree is essential. The maturity and sophisticated understanding of a subject area that is the typical result of a doctoral study program, plus the discipline and depth of understanding stemming from the dissertation experience, produce a new, vital plane of understanding in the individual. In addition, it has often been correctly said that one of the major benefits derived from a doctoral program is that one has learned how to teach oneself. This ability is exactly what is required if one is to engage in research, because research by definition is teaching oneself about a topic concerning which little is known.

It was a widespread practice, until after World War II, for engineering schools to employ faculty with master's degrees, but criticism by national study committees fell upon much of U.S. engineering education because faculty members in many schools did not possess a sufficiently advanced level of education in science and mathematics to prepare their students for modern technology.[3,4] Recommendations were made that a major upgrading in educational level for engineering faculty was needed. Today, with the few exceptions previously noted, engineering schools with graduate and research programs almost universally require new faculty to have doctor's degrees.

Adjunct and Part-Time Faculty

As mentioned in the previous section, many universities historically have used engineers from industry in the roles of adjunct or part-time faculty. Such individuals are usually chosen for their expertise in some branch of engineering practice, and frequently they can offer material to both graduate and undergraduate students that is not within the experience of the regular faculty. This enriches the students' education, though judicious selection of appropriate individuals is necessary, with special emphasis on teaching abilities. It is also essential for such individuals to be willing to devote much time to a task that, even though it is labeled "part-time," is likely to be much more demanding than was initially perceived. While many part-time instructors have been very successful and are highly prized by their academic colleagues, others have not been successful, often because they underestimated the time and commitment required to organize and teach courses of high quality.

Universities can ill afford to rely extensively on adjunct or part-time instructors. During the initial stages of the faculty shortages of the early 1980s, some schools relied so extensively on part-time faculty that they later came to believe that the quality of their instruction had declined as a consequence. As was pointed out earlier, much education occurs outside of the classroom, and some of the most important of that depends on unstructured student–faculty contact. Put succinctly, the faculty—at least most of them—must be on hand and available. As a general rule, part-time instructors are less available than regular faculty for the kind of frequent student–faculty interaction that is desirable. Although there are exceptions, of course, most part-time instructors are on campus for scheduled classtime plus a nominal number of office hours for consultation, and then must hurry back to their regular jobs. After all, they generally have full-time responsibilities elsewhere, and must take those responsibilities seriously. Also, part-time instructors rarely are on hand enough to participate in program development, thus actually throwing an additional burden on the regular faculty. If a majority of the faculty is part time, the result can even be a loss of control of the curriculum. As a final matter of concern, many universities are simply not located in regions with enough industry to make many part-time instructors available.

In spite of the disadvantages, adjunct and part-time faculty can make a positive contribution to an engineering program, particularly when they can make a contribution by representing true-to-life professional problem solving. Therefore, colleges and universities should be encouraged to use adjunct and part-time faculty for the purpose of enriching the students' educational graduate and undergraduate programs. However, excessive reliance on part-time faculty can impair the quality of an educational program, and it requires careful monitoring.

Faculty Development and Self-Renewal

The need for continuing education and self-renewal for engineers was discussed in an earlier section, but faculty need self-renewal, too. Continual self-renewal is a part of the definition of a scholar, and universities expect their engineering faculty members to be scholars. This is the source of the common requirement that faculty members engage in productive research programs, because persons involved in such programs are constantly working at the frontiers of their fields and must keep current in those areas.

The requirement for research is customary in institutions referred to as research universities, and the teaching loads in those institutions are generally adjusted to a level that makes a research commitment possible. In other institutions—those with little or no research involvement and with high teaching loads—the need for faculty self-renewal becomes critical.

For engineering faculty, a special kind of self-development arises from the need for a close association with professional practice. This is true to a lesser degree, or not at all, for professors of physics, chemistry, and mathematics. In those fields, a professional—meaning a person educated to the doctoral level—does the same sorts of things that the faculty do. But most engineering graduates enter career activities that differ markedly from those of most of the faculty. Hence, engineering faculty must keep in touch with industrial practice. Furthermore, in some fields the pace of engineering research is being set by industry, not by academia. This situation makes it all the more imperative for faculty in those fields to maintain industrial contact.

Consulting relationships represent one of the most common ways of maintaining contact between faculty and professional practice, and this constitutes one of the strongest justifications for consulting. Industry–university joint research programs are another way to maintain contact, although this particular means may be available principally to research universities, whereas the opportunity to consult is not limited to research universities. (Consulting and other industry–university interactions are discussed below.)

Sabbatical leaves are administrative devices expressly intended for faculty self-renewal, although opportunities for sabbaticals are not uniformly available throughout all educational institutions. Certainly, the wide availability of sabbatical leaves for purposes of self-renewal is strongly to be encouraged, with stipends to the faculty set at a reasonably large fraction of the regular salary so that the leaves are financially possible.

Summer jobs in industry represent an attractive way to maintain professional contact, although such jobs obviously have to be meaningful enough to serve the intended purpose. However, industrial summer jobs for faculty can be counterproductive for faculty in research universities, because summer is often the only time they can devote intensive effort to their research projects, the rest of the year being fragmented by teaching assignments and administrative duties.

In the context of self-renewal, a question arises concerning teaching loads. In recognition of the need for constant self-renewal through research, faculty in research universities carry lower teaching loads than do faculty in nonresearch universities. Yet, even in research universities there is concern among faculty that teaching loads are too heavy, considering other responsibilities. (See Tables 25 through 28 in

Chapter 9,"University–Industry Interactions," for further information on how faculty members spend their time.) There are serious questions as to whether faculty members in institutions where they must spend the majority of their time lecturing to classes can have enough time for self-renewal in view of other instruction-related assignments such as preparation for class, counseling students, and grading papers. A reappraisal of teaching loads would appear to be in order in such cases, with obvious implications for budgetary support.

Findings and Recommendations

1. Schools that emphasize undergraduate education can utilize many faculty who do not have doctor's degrees. Universities with significant graduate and research programs will also be able to utilize some nondoctoral faculty in special roles. However, the faculty in such "research universities" should, in the overwhelming majority, have doctor's degrees.

2. Colleges and universities should be encouraged to use adjunct and part-time faculty to enrich students' educational programs. However, excessive reliance on part-time faculty can impair the quality of an educational program and should be avoided.

3. Faculty members need opportunities for self-development and self-renewal, just as do engineers in industry. Methods for accomplishing such self-renewal include participation in research, industrial consulting, sabbatical leaves, and summer jobs in industry.

9
University–Industry Interactions

It has frequently been suggested in recent years that a major increase in university–industry engineering research should take place in order to enhance the financial base for universities and to help support the needed increase in doctoral enrollments. A study of such relationships, entitled *University–Industry Research Relationships*, was published by the National Science Board (NSB) in 1982.[57,58] In an investigation of views held by universities and companies, the following were found to be the principal motives for increasing university–industry research:

Industrial motivations:

1. To obtain access to manpower (students and professors).
2. To obtain a window on science and technology.
3. To solve a problem or to get specific information unavailable elsewhere.
4. To obtain prestige or enhance the company's image.
5. To make use of an economical resource.
6. To provide general support of technical excellence.
7. To be good local citizens or to foster good community relations.
8. To gain access to university facilities.

University motivations:

1. Industry provides a new source of money, which helps diversify the university's funding base.
2. Industrial money involves less red tape than government

money, and the reporting requirements are not as time consuming.
3. Industrially sponsored research provides student exposure to real-world research problems.
4. Industrially sponsored research provides a chance to work on an intellectually challenging research program that may be of immediate importance to society.
5. Currently, some government funds are available for applied research, based on a joint effort between university and industry.
6. Such research will provide better training for the increasing number of graduates going to industry.

The National Science Board study also found that only a small fraction of university research money comes from industry. In 1981 the funding sources were as follows:

R&D in Universities and Colleges, 1981

Source of Funds	($ million)	Percent
Federal government	4,100	65.0
Industry	240	3.8
Universities and colleges	1,490	23.6
Other	480	7.6
	6,310	100.0

However, these figures represent *all* research funding for all schools and colleges within the universities. For engineering schools, the percent coming from industry is substantially higher than 3.8 percent, as was shown in Table 24. For the 30 universities in Table 24, the percentage coming from industry was approximately 15 percent.

The predominant view of those surveyed in the NSB study[57] is that the major source of research funding in universities will continue to be the federal government. Some industrial respondents in that study expressed the opinion that the amount of industrial funding could be increased but that it could not be expected to fill any large funding drop by the federal government. In particular, the following caution was sounded:

Companies do intend to draw more direct ties to universities, but resources are limited, and they already support the university research endeavor through taxes. It is important to remember that industry has to pursue a direction which strengthens its own long-term interests, and the university must pursue a direction based on its function in society. These directions can intersect but to a limited extent. Only the government has the resources and network capabilities to monitor the complex U.S. research system and ensure that we have a broad technical base. [Ref. 57]

Nevertheless, universities and industries are showing increasing interest in developing closer relationships, and many companies that have longstanding relationships with academia have demonstrated their sensitivity to the different purposes of these two societal institutions and have found ways to make their associations mutually fruitful.

A major way that universities and industries can benefit through closer ties is by the development of groups sometimes known as advisory councils or visiting committees. Such groups can provide two-way communication, helping universities to evolve their programs in concert with industrial needs and helping to keep industry in touch with what is going on in academia. Frequently, instead of reporting to a dean of engineering, such a group is appointed by and is advisory to the president or chancellor of an institution. In such cases the dean meets regularly with the committee, but when the committee speaks it is with an independent voice, directly to the top officer. A committee of this type not only can be a powerful and beneficial voice within an academic institution, it can also speak with competence and authority in public forums, for example, to state legislatures and to Congress.

This report will not undertake to review all of the information published in *University-Industry Research Relationships* but will underscore certain potential problem areas together with possible directions that might be taken. The problem areas are these: (1) the appropriate nature of industrially sponsored research, (2) patent problems, (3) consulting relationships, and (4) conflicts of interest.

The Appropriate Nature of Industrially Sponsored Research

Industrial organizations, by their very nature, are in the business of developing products or services that can be sold at a profit. Universities, on the other hand, are in the business of producing educated people and basic knowledge. Industrial organizations need to develop proprietary positions on new products or processes, and often they must impose conditions of confidentiality on their research and development activities. But universities are in the opposite position: by virtue of the nature of universities, the knowledge they develop should be freely available to all. Public universities especially are often required by legislative action to operate under "fishbowl" conditions, with the knowledge they develop made available for the public good and not for a select few. Thus, there is an automatic mismatch of purposes between industrial organizations and universities, and these purposes must be brought into reasonable harmony if productive collaboration is to occur.

As a general rule, the closer a university comes to the activity of product development, the less likely it is that the purposes of the university will be well served. Product development can be highly creative, but it is a process of specialization—bringing knowledge to a focus on a specific item or service. University faculties and students serve their publics best when they produce information that is broadly generalizable. In this manner they can influence the development of their own fields and make contributions of greater potential value to the public welfare than if they engage in the kind of focusing activity that emphasizes a particular product. The process of product development usually involves only a small proportion of activity that is generalizable in nature. A proportionally large amount of time must be devoted to solving the difficult problems that arise in reducing an idea to practice but which are by their nature low in generalizable content.

The welfare of the universities' students is at stake if they are extensively engaged in product development activities while they are still students. While it is certainly true that exposure to product development is valuable for an engineering student, it is questionable whether a large time commitment to the "reduction-to-practice" phase is an appropriate use of the short time that a student, whether undergraduate or graduate, is in the university setting. Such a time commitment can crowd out the opportunity to acquire a wide spectrum of fundamental and transferable knowledge that a student may need throughout his or her career. If the further condition of secrecy is imposed, then one of the principal advantages of an academic setting is lost, namely, the opportunity to share knowledge and to learn from one's colleagues. Finally, there is the possibility that students may unknowingly be bent to the hidden purposes of others if proprietary and potentially profitable products are in view.

Because of the foregoing considerations, industrially sponsored research in universities should be free of secrecy constraints and should be as general as possible so that the learning experiences of the students are transferable to a variety of future needs. An exception to the matter of secrecy might be made for limited periods of time to permit the filing of patent applications, but other than that, the imposition of secrecy can limit a student's opportunity to gain the most from his or her educational experience. In a similar manner, an industrially sponsored project, to be appropriate for universities, should also be free from potential conflicts of interest on the part of the faculty principal investigator. If the investigator, for example, is also an officer of the firm sponsoring the research or has a substantial interest in the firm, the course of the project may be directed towards the firm's interests rather than those of the student. Further, the incentive for secrecy in such a case is increased on the part of the faculty investigator regardless of any formal contract provisions against secrecy, because the proprietary position of his or her firm can be enhanced through secrecy. Such barriers between students and other students or faculty can inhibit a student's learning experiences.

Patents

The ownership of so-called intellectual property—usually meaning patents and copyrights—is a vexing problem for universities, particularly as industry increasingly enters the picture.

When a company funds all or nearly all of a project that results in a patentable device or process, the industrial sponsor naturally feels that it should have some privileged position vis-à-vis that patent. Yet some universities take the position that all ownership rights should be vested in the university and that the sponsor should pay for a royalty-bearing license if it wishes to use the patent. Public interest groups have sometimes insisted that publicly supported universities should adopt a policy of total ownership by the universities, because to grant ownership to private businesses, they say, would subvert public resources to private purposes. In support of their view, they point out that industrial sponsors rarely pay for *all* the costs of university projects and that efforts to impose overhead charges by universities are often met with resistance. It is also asserted that students who come to a university for an education should not find themselves in a position of serving private interests.

These issues are heavily loaded with value judgments and political philosophies; however, if meaningful industry–university partnerships are to be developed, resolutions must be found. On the part of industry, there must be a readiness to pay full costs, including overhead costs, if the sponsor expects to have a privileged position regarding ensuing patents or copyrights. On the part of the university, there should be a willingness to grant ownership of the patent to such a full sponsor, or—which is virtually the same thing—to grant a royaltyfree exclusive license. If the sponsor provides less than full funding, then some appropriately scaled lesser rights should be granted. If full funding is provided by the university, then full ownership of any resulting patent rights should be with the university.

Universities have become somewhat more aggressive in seeking patent income than in the past, sometimes acting under urging from their trustees that they should function more as income-producing enterprises. In some cases the urgings are accompanied by examples of cases in which the incomes have been quite large. Yet the record shows that for most universities the income from patents is small. In the NSB study mentioned above, information on patent income was sought from the 36 universities believed to be the most successful in gaining patent income. The responses gave the following information:[57]

Frequency Table: Total Patent Royalties Received by Sample of Universities—1980 and 1981

Gross Income	Frequency 1980	Frequency 1981
0–$ 99,999	10	7
$100,000–$199,999	3	4
$200,000–$299,999	3	2
$300,000–$399,999	3	0
$400,000–$499,999	0	1
Over $500,000	6	8
	25	22

Thus, half of the responding universities in 1981 had royalty incomes of less than $200,000 each, and only eight had incomes exceeding $500,000. From these incomes must be deducted the costs of administering the patent programs, including the legal costs incurred in the patenting process, and the costs of marketing licenses. Even though the income may be welcome to the sponsoring universities, it is very small in comparison to their overall programs.

On the other hand, income accruing to the individual inventors—who are usually full-time employees of the university—has occasionally been significant, sometimes reaching into the tens of thousands of dollars. A legitimate question arises concerning how much additional income should be given to a full-time university employee who produces a lucrative patent, and whether the reward structure of the university is unduly skewed in the process. Most university employees do not work in areas where patents are even possible. Faculty in the humanities and social sciences—where the question of patents is generally irrelevant—might well argue that their contributions have more long-range value to society than does a patent whose lifetime is limited. Even within the engineering school many faculty work on very fundamental topics. If unduly large amounts of extra income go to those who produce patents, there may be an unwanted drawing away from work in fundamental, unpatentable fields toward areas where patents might result. This could produce an unfortunate effect on university research if it resulted in a drift away from fundamental areas of research in which universities are expected to excel.

Nevertheless, universities usually provide for sharing patent income with the inventors. The National Science Board study shows that the sharing typically ranges from 15 percent to 50 percent of net royalty income, sometimes with a sliding scale that provides a decreasing percentage as the royalty income increases.[57] For full-time employees, it could be argued that the inventor has already received a salary for the activity that produced the patent and that the major share of royalty income should go to reimburse the institution that paid the salary. Of course, such a consideration would not apply if the inventor is a student who is not a paid employee in the activity that produced a patent.

Somewhat similar considerations apply in the case of other intellectual activities, the types usually protected by copyright. Prominent among these are textbooks, videotapes, and computer programs. The same general principle described for patents should apply to all of these: if the activity is funded as a work assignment by the university, then ownership and income should belong to the university, with the author perhaps receiving a nominal award. If the activity is conducted substantially outside of university auspices, the university should seek no rights.

Consulting

It is usually held that participation in consulting work is healthy for the faculty of a professional school such as engineering. A 1978 survey showed that 62 percent of engineering faculty engaged in paid consulting of some kind, the most for any field.[59] Consulting activity is an important way for faculty to maintain contact with professional practice and it helps them stay at the cutting edge of technology. Otherwise, school course work could become increasingly abstract and analytical and fail to provide the appropriate touchstone with current practice that is vital in a high-quality educational program. It is also the case that faculty consulting sometimes results in research contracts that are placed with the universities, thus enhancing the educational opportunities for students. In fact, it is difficult to envision substantial increases in industry–university research cooperation without the enabling influence of direct faculty contacts of the kind represented by consulting.

In spite of these benefits, consulting by faculty members often comes

TABLE 25 Professional Activity of Engineering Faculty, by Type of Institution: 1978–1979 (mean hours per week)

	Doctorate Institutions	Nondoctorate Institutions
All activities	49.1	46.0
Instructional activities	15.5	23.2
Research	14.7	3.2
Public service and administration	10.0	6.0
Total outside income-producing activities	5.8	10.3
Consulting	(3.3)[a]	(4.8)[a]
Publication	(1.7)[a]	(4.4)[a]
Other	(0.8)[a]	(1.2)[a]
Continuing education and professional enrichment	3.0	3.2

[a]Numbers in parentheses are breakdowns of the total outside income-producing activities.
SOURCE: *Activities of Science and Engineering Faculty in Universities and 4-year Colleges: 1978–1979* (Washington, D.C.: National Science Foundation, NSF 81–323).

under fire, and it is appropriate to consider the reasons for this criticism. One source of criticism stems from the fact that most people react adversely to the image of a professor who, instead of teaching classes, is off doing consulting and perhaps reaping enormous financial rewards in the process. Sometimes the reaction to this image is to conclude that professors in professional schools do not need to be paid competitive salaries because of the presumed outside income. However, the data shown in Table 25, taken from a special survey conducted by the National Science Foundation,[59] do not support this image. The first thing that most people find surprising in such surveys is the length of the workweek of the average faculty member. Table 25 shows that the workweek for faculty in Ph.D.-granting universities averaged 49.1 hours, and in non-Ph.D.-granting universities, 46.0 hours.* Of the total workweek in Ph.D.-granting universities, an average of only 3.3 hours was spent in consulting, and 2.5 hours in other outside activities, such as textbook writing. Oddly, and again contrary to popular impression, faculty in non-Ph.D.-granting institutions spend even more time in outside activities: 4.8 hours per week in consulting and 5.6 hours in other activities. Table 25 also shows some results that are *not* surpris-

---

* A 1982–1983 survey of faculty in the University of California showed an average workweek of 60.6 hours in university-related activities, divided as follows: instruction—27.5 hours; research—23.9 hours; university service—4.8 hours; professional activities and public service—4.4 hours. No data were included on consulting.

ing: faculty in doctorate institutions spend less time in instructional activities than do those in nondoctorate institutions (15.5 hours versus 23.2 hours) and more time in research (14.7 hours versus 3.2 hours).

Tables 26, 27, and 28 show other comparisons between groups of faculty. Table 26 shows that engineering faculty who possessed doctoral degrees had a mean workweek of 50.0 hours, and those without doctorates had a mean workweek of 41.4 hours. Doctorate faculty spent less time in instructional activities and more time in research than did nondoctorate faculty, which is in accord with common conception. Contrary to the usual impression, however, doctorate faculty spent less time in consulting (3.4 hours) than did nondoctorate faculty (4.9 hours).

Table 27 compares engineering faculty against science faculty at doctorate institutions. Engineering faculty in general engage in more instructional activities and in less research than do other groups. (Note that these relationships are reversed when the comparison is with social scientists.) Engineers also consult more.

Table 28 deals with nondoctorate institutions, showing that engineering faculty in these institutions have significantly longer workweeks than other groups. The time of engineering faculty spent in instructional activities is greater than that of faculty in the life sciences and social sciences but slightly less than that spent by faculty in the physical sciences. Engineering faculty time spent in consulting (4.8

TABLE 26 Professional Activity of Engineering Faculty, by Possession of Doctor's Degree: 1978–1979 (mean hours per week)

	Doctorate Faculty	Nondoctorate Faculty
All activities	50.0	41.4
Instructional activities	16.9	21.2
Research	14.0	1.5
Public service and administration	9.0	7.8
Total outside income-producing activities	7.0	8.0
Consulting	(3.4)[a]	(4.9)[a]
Publication	(3.0)[a]	(1.3)[a]
Other	(0.6)[a]	(1.9)[a]
Continuing education and professional enrichment	3.1	2.8

[a]Numbers in parentheses are breakdowns of the total outside income-producing activities.
SOURCE: *Activities of Science and Engineering Faculty in Universities and 4-year Colleges: 1978–1979* (Washington, D.C.: National Science Foundation, NSF 81-323).

TABLE 27 Professional Activity of Science and Engineering Faculty for Doctorate Institutions, by Field: 1978–1979 (mean hours per week)

	Engineering	Life Sciences	Physical Sciences	Social Sciences
All activities	49.1	50.6	49.6	48.1
Instructional activities	15.5	13.4	13.8	18.2
Research	14.7	18.8	20.5	10.5
Public service and administration	10.0	11.0	8.0	8.4
Total outside income-producing activities	5.8	3.6	3.9	4.8
Consulting	(3.3)[a]	(0.8)[a]	(1.1)[a]	(0.9)[a]
Publication	(1.7)[a]	(2.1)[a]	(2.5)[a]	(3.6)[a]
Other	(0.8)[a]	(0.7)[a]	(0.3)[a]	(0.3)[a]
Continuing education and professional enrichment	3.0	3.9	3.4	6.1

[a]Numbers in parentheses are breakdowns of the total outside income-producing activities.
SOURCE: *Activities of Science and Engineering Faculty in Universities and 4-year Colleges: 1978–1979* (Washington, D.C.: National Science Foundation, NSF 81-323).

TABLE 28 Professional Activity of Science and Engineering Faculty for Nondoctorate Institutions, by Field: 1978–1979 (mean hours per week)

	Engineering	Life Sciences	Physical Sciences	Social Sciences
All activities	46.0	43.2	42.1	40.6
Instructional activities	23.2	21.0	23.9	19.4
Research	3.2	6.5	3.8	5.9
Public service and administration	6.0	8.5	6.8	6.5
Total outside income-producing activities	10.3	1.7	2.7	2.1
Consulting	(4.8)[a]	(0.3)[a]	(0.5)[a]	(0.5)[a]
Publication	(4.4)[a]	(0.6)[a]	(1.6)[a]	(1.4)[a]
Other	(1.2)[a]	(0.9)[a]	(0.6)[a]	(0.2)[a]
Continuing education and professional enrichment	3.2	5.5	4.9	6.7

[a]Numbers in parentheses are breakdowns of the total outside income-producing activities.
SOURCE: *Activities of Science and Engineering Faculty in Universities and 4-year Colleges: 1978–1979* (Washington, D.C.: National Science Foundation, NSF 81-323).

hours per week) is substantially higher than for the other groups (0.3 to 0.5 hours per week).

The general conclusions to be drawn from these tables is that engineering faculty members have a long workweek and spend less of it in outside activities than is generally thought. In spite of this reassurance, there have been cases in which faculty consulting has been excessive, leading to neglect of university duties. University administrators need to be alert to such transgressions and certainly should not tolerate them. Most universities have policies that permit and even encourage consulting, usually limited to one day per week, but also require that all of a faculty member's responsibilities to the university simultaneously must be fulfilled. A faculty member should regard teaching as a prime responsibility; it is hard to see how this responsibility is being met properly if a professor is consulting when a class is scheduled, perhaps leaving a teaching assistant to meet with the class. But teaching is not the only such responsibility. A professor's graduate students have a major claim on his or her time, even though such time may not be formally scheduled. The institution also has a claim on the faculty member, for such things as institutional governance, curriculum development, and peer faculty review. Finally, in a research university, there is a major expectation that a faculty member will initiate and lead significant research projects. In the aggregate, these activities constitute what is meant by "meeting one's responsibilities to the university." Even though it is admittedly sometimes difficult to form appropriate value judgments regarding whether an individual is fulfilling his or her responsibilities, the task of doing so cannot be avoided. Consulting is just one of the many elements that must go into the total evaluation of a professor's contributions to the institution.

Another source of criticism of faculty consulting is that the faculty member is competing against regular consulting firms but operating from a sheltered position. The argument is that a faculty member is protected by receiving a full-time salary and thus can unfairly compete with a regular consulting firm that has many overhead costs to bear.

There seems to be little justification for faculty members' placing themselves in this kind of position. If the practice of unfairly competing with regular firms is carried to excess, faculty could even find themselves the targets of unwelcome legislative action. Furthermore, most of the kinds of projects that would be involved in such competition would probably be relatively routine and repetitive. If such projects are repetitive, it is unlikely that the faculty member could successfully argue that the consulting activity is at the cutting edge, and thus of maximum benefit to his or her teaching.

A final criticism of faculty consulting is political in nature. Some groups have insisted that faculty consulting should be prohibited altogether, at least from public universities. To do otherwise, they claim, is to divert public resources to private interests and to work to the disadvantage of society. They have particularly directed their attention to consulting for petroleum and chemical firms, the nuclear power industry, and defense firms. In California, court action has been brought against the University of California, because, it is claimed, the activities of the university's huge Agricultural Experiment Station have served principally to benefit rich farmers, to the detriment of other segments of society. In the case of the experiment station, faculty consulting is not specifically at issue, because outside consulting activity is prohibited for faculty associated with the station, but the overall issue is the same: it is claimed that special influences, backed by money, are brought to bear on the university to divert its attention in ways that will benefit those special interests and disadvantage others.

There is no easy answer to such charges. Different views of the varying advantages and disadvantages will always be a function of personal value judgments and political opinions regarding the fashion in which society should be structured. However, it is generally agreed that faculty consulting is beneficial to students, to universities, and to society, provided it meets the conditions discussed here. Industry–university research also provides benefits to all, again provided it meets the conditions outlined. If universities were to sever all relationships with industry, as advocated by some, they would tend toward a closed-in, "ivory tower" character even more than they do now. Universities are already criticized on that account in some quarters, and further movement in that direction is hardly to be sought.

Some universities require their full-time faculty members to report their outside activities, and proposals have been made that they should also be required to report the specific identities of the organizations for which they consult, the amount of time spent, and the income derived. It would seem that no invasion of privacy is involved in asking faculty to report the names of the organizations for whom they consult, and the amount of time spent. After all, they are *full-time* employees of their universities, and their employers would surely be within their rights in asking for such information. Also, the public interest would be served by having this information available, although there might well be different rules for public and private universities. However, the reporting of outside income *does* seem to be an invasion of privacy, since it provides no information relating to the employers' interests that cannot be gained from a full reporting of time spent. Income information, even though it may be collected by universities on a confidential basis, could be forced into the public domain by use of the Freedom of Information Act, and could be used for the unwarranted harassment of individuals.

Faculty Conflicts of Interest

The discussion so far raises the question of the kinds of faculty involvement with industry that are appropriate and of those that are inappropriate. It has already been asserted that faculty consulting is desirable, if controlled, and the same has been said about industry–university research cooperation. However, recent dramatic developments in the fields of computers and genetic engineering have raised the question of faculty codes of conduct to prevent conflicts of interest.

It would seem inappropriate for a faculty member to be a principal investigator within a university on a project that is funded by a company for which the faculty member simultaneously is an officer, a director, or a significant owner (owning, say, 5 percent or more of the company). The prospect of conflicts of interest arising are too great to be tolerated in such a situation. Even if a faculty member successfully avoided actual improprieties in such a case, the suspicions on the part of other faculty and students could be so great that the welfare of the educational enterprise could be damaged.

The question of service simultaneously as a principal investigator on an industry-funded project and as a consultant for the same company is somewhat more clouded. Some have proposed that such relationships should be prohibited, but it should be noted that industry–university research contracts often would not come to pass if it were not for preexisting consulting relationships. To prohibit such connections would be to forfeit major benefits in exchange for a questionable gain. Such relationships should be reported and subjected to regular review by the university administration but should not be prohibited.

Findings and Recommendations

1. Closer ties between industry and engineering education should be fostered. Such ties can take many forms, such as increased industrially sponsored research; faculty consulting; industry financial support of graduate fellowships, modern equipment, facilities, and departmental expense. An especially beneficial form of industry–university cooperation is the establishment of industrial advisory councils to provide input to campus administrations and governmental bodies.

2. Industrially sponsored research in universities should be free of secrecy constraints and should be as general as possible so that the learning experiences of the students can be transferred to a variety of future needs.

3. With respect to patents and other intellectual property, industrial sponsors should be prepared to pay full costs, including overhead, if they expect to have special rights with respect to such property. In such cases of full cost sponsorship, universities should grant ownership of the patent or other intellectual property to the sponsor, or grant a royaltyfree exclusive license.

4. Outside consulting by faculty members should be encouraged, provided it supports and helps improve the academic programs of the university. Consulting of a routine sort should be discouraged, as should consulting that is competitive with regular consulting organizations. However, full-time faculty members should limit their consulting and other outside activities so that university responsibilities are not interfered with.

5. Faculty members must scrupulously avoid conflicts of interest. It is inappropriate for a faculty member to be a principal investigator within a university on a project funded by a company for which the faculty member is simultaneously an officer, a director, or a significant owner.

Notes

1. "Longhairs and Short Waves," *Fortune*, November 1945, p. 163.
2. Daniel C. Drucker, "Engineering—Key to Industrial Leadership," speech at Case Western Reserve University, April 14, 1982; L. Lessing, "M.I.T. and the New Breed of Hairy Ears," *Fortune*, February 1961, p. 129.
3. L. E. Grinter (Chairman), *Report on Evaluation of Engineering Education* (Washington, D.C.: American Society for Engineering Education, June 15, 1955).
4. J. B. Wiesner (Chairman), *Meeting Manpower Needs in Science and Technology. Report Number One: Graduate Training in Engineering, Mathematics, and Physical Sciences* (Washington, D.C.: President's Science Advisory Committee, The White House, December 12, 1962).
5. E. A. Walker (Chairman), *Goals of Engineering Education: Final Report of the Goals Committee* (Washington, D.C.: American Society for Engineering Education, January 1968).
6. *New York Times*, Nov. 1, 1970, Sec. VI, p. 28.
7. *New York Times*, Nov. 7, 1970, p. 12; Nov. 8, 1970, Sec. I, p. 1; Nov. 20, 1970, Sec. III, p. 3; May 18, 1971, p. 1; June 13, 1971, p. 86.
8. *New York Times*, Dec. 20, 1970, p. 29; March 29, 1971, p. 35.
9. *New York Times*, Jan. 31, 1971, Sec. III, p. 3.
10. *New York Times*, Jan. 4, 1971, p. 30; Jan. 21, 1971, Sec. III, p. 13; July 5, 1971, p. 31.
11. *San Francisco Examiner*, Aug. 1, 1981.
12. John D. Alden, "What is Happening in Engineering?" Presentation at the Annual Conference, American Society for Engineering Education, June 21, 1972.
13. *Science and Engineering Doctorate Supply and Utilization, 1968-80*, NSF 69-37 (Washington, D.C.: National Science Foundation, 1969).
14. *1969 and 1980 Science and Engineering Doctorate Supply and Utilization*, NSF 71-20 (Washington, D.C.: National Science Foundation, May 1971).
15. *Summary Report 1982: Doctorate Recipients From United States Universities* (Washington, D.C.: National Academy Press, 1983).

NOTES

16. *Science and Engineering Doctorates: 1960-81*, NSF 83-309 (Washington, D.C.: National Science Foundation, 1983).
17. *Summary Report 1983: Doctorate Recipients From United States Universities* (Washington, D.C.: National Academy Press, 1984).
18. *Science Indicators, 1980* (Washington, D.C.: National Science Board, 1981).
19. *Science and Engineering Education: Data and Information*, NSF 82-30 (Washington, D.C.: National Science Foundation, 1982).
20. J. Geils, "The Faculty Shortage: A Review of the 1981 AAES/ASEE Survey," *Engineering Education*, November 1982, pp. 147-158.
21. W. Edward Lear, "The State of Engineering Education," *Journal of Metals*, February 1983, pp. 48-51.
22. *The Quality of Engineering Education* (Washington, D.C.: National Association of State Universities and Land-Grant Colleges, November 7, 1982).
23. J. Geils, "The Faculty Shortage: The 1982 Survey," *Engineering Education*, October 1983, pp. 47-53.
24. *Projected Employment Scenarios Show Possible Shortages in Some Engineering and Computer Specialties*, NSF 83-307 (Washington, D.C.: National Science Foundation, February 23, 1983).
25. *Energy-Related Manpower, 1983* (Oak Ridge, Tenn.: U.S. Department of Energy, November 1983).
26. L. V. Baldwin and K. S. Down, *Educational Technology in Engineering* (Washington, D.C.: National Academy Press, 1981).
27. Lionel V. Baldwin, "Organizing and Managing a Technology-based National Engineering College." Paper presented at the Sixth National Conference on Communications Technology in Education and Training, Washington, D.C., February 22-24, 1984.
28. *Engineering and Technology Enrollments, Fall 1981* (New York: Engineering Manpower Commission, 1982).
29. Sue E. Berryman, *Who Will do Science?* (New York: The Rockefeller Foundation, November 1983).
30. *Proceedings of the Engineering Educators Workshop*, Society of Women Engineers National Convention, June 29, 1979, San Francisco.
31. *Women in Engineering—Beyond Recruitment*, Proceedings of the Conference, Cornell University, Ithaca, N.Y., June 22-25, 1975.
32. B. M. Vetter, E. L. Babco, and S. Jensen-Fisher, *Professional Women and Minorities: A Manpower Data Resource Service*, 3rd ed. (Washington, D.C.: Scientific Manpower Commission, April 1982).
33. R. M. Hall and B. R. Sandler, "Women Winners," in *Project on the Status and Education of Women* (Washington, D.C.: Association of American Colleges, 1982).
34. V. F. Nieva and B. E. Gutek, "Sex Effects on Evaluation," *The Academy of Management Review*, vol. 5, no. 2, pp. 167-176, 1980.
35. B. Rosen and T. H. Jardee, "Influence of Sex Role Stereotypes on Personnel Decisions," *Journal of Applied Psychology*, vol. 59, pp. 9-14, February 1974.
36. B. Rosen and T. H. Jardee, "Sex Stereotyping in the Executive Suite," *Harvard Business Review*, vol. 52, March-April, pp. 45-58, 1974.
37. B. Rosen and T. H. Jardee, "Effect of Applicant's Sex and Difficulty of Job on Evaluations of Candidates for Managerial Position," *Journal of Applied Psychology*, vol. 59, August, pp. 511-512, 1974.
38. M. Patterson and L. Sells, "Women Dropouts from Higher Education," in A. Rossi

and A. Calderwood, eds., *Academic Women on the Move* (New York: Russell Sage Foundation, 1973), pp. 88-89.
39. C. L. Attwood, *Women and Fellowship and Training Programs: Project on the Status and Education of Women* (Washington, D.C.: Association of American Colleges, 1972), App. B, pp. 20-24.
40. *Women and Minorities in Science and Engineering* (Washington, D.C.: National Science Foundation, January 1982).
41. *Climbing the Academic Ladder: Doctoral Women Scientists in Academe* (Washington, D.C.: National Academy of Sciences, 1979).
42. J. R. Cole, "Women in Science," *American Scientist*, vol. 69, no. 4, pp. 385-391, July-August 1981.
43. "Women EEs Reported to Earn $2600 Less than Male Peers," *The Institute*, vol. 8, no. 3, p. 1, 1984.
44. P. Kraft and S. Dubnoff, "Software Workers Survey," *Computerworld*, November 14, 1983, pp. 4-16.
45. *The Master's Degree: A Policy Statement* (Washington, D.C.: The Council of Graduate Schools in the United States, December 1976).
46. *Manual of Graduate Study in Engineering* (Washington, D.C.: American Society for Engineering Education, 1952).
47. *Engineering College Research and Graduate Study*, *Engineering Education*, March 1983.
48. Paul Doigan, "Engineering Degrees Granted, 1983," *Engineering Education*, April 1984, pp. 640-645.
49. C. M. Van Atta, W. D. Decker, and T. Wilson, *Professional Personnel Policies and Practices of R&D Organizations* (Livermore, Calif.: Lawrence Livermore Laboratory, December 6, 1971).
50. G. W. Dalton and P. H. Thompson, "Accelerating Obsolescence of Older Engineers," *Harvard Business Review*, September-October 1971, pp. 57-67.
51. W. K. LeBold, R. Perruci, and W. Howland, "The Engineer in Industry and Government," *Engineering Education*, March 1966, pp. 237-273.
52. W. K. LeBold, K. W. Linden, C. M. Jagacinski, and K. D. Shell, *National Engineering Career Development Study: Engineers' Profiles of the Eighties* (West Lafayette, Ind., Purdue University, June 1983).
53. *The Doctor of Philosophy Degree: A Policy Statement* (Washington, D.C.: The Council of Graduate Schools in the United States, October 1977).
54. John D. Kemper, "Post-Master's Engineering Programs Directed to Industry's Needs," Paper presented at the 86th Annual Conference, American Society for Engineering Education, University of British Columbia, June 19-22, 1978.
55. Lawrence N. Canjar, "Examples of Professional Schools of Engineering," *Engineering Education*, February 1972, p. 442.
56. L. E. Grinter, "Defining a Professional School of Engineering," *Engineering Education*, January 1975, p. 279.
57. *University-Industry Research Relationships: Selected Studies*, NSB 82-2 (Washington, D.C.: National Science Board, 1982).
58. *University-Industry Research Relationships: Myths, Realities, and Potentials*, NSB 82-1 (Washington, D.C.: National Science Board, 1982).
59. *Activities of Science and Engineering Faculty in Universities and 4-year Colleges: 1978/79*, NSF 81-323 (Washington, D.C.: National Science Foundation, December 1981).

NOTES

60. *Engineering College Research and Graduate Study, Engineering Education,* March 1984.
61. *Signs of Trouble and Erosion: A Report on Graduate Education in America* (New York: New York University, December 1983).
62. *Discussion Issues, 1983: The Engineering Mission of NSF Over the Next Decade,* Vol. II (Washington, D.C.: National Science Board, June 1983), p. 122.
63. *Discussion Issues, 1983: The Engineering Mission of NSF Over the Next Decade,* Vol. I (Washington, D.C.: National Science Board, June 1983), p. 40.
64. *Projected Response of the Science, Engineering, and Technical Labor Market to Defense and Nondefense Needs: 1982-87,* NSF 84-304 (Washington, D.C.: National Science Foundation, January 1984).
65. P. Doigan, "ASEE Survey of Engineering Faculty and Graduate Students, Fall 1983," *Engineering Education,* October 1984.
66. *Research Laboratory of Electronics: 1940 + 20* (Cambridge, Mass.: MIT, May 1966).